Animal Virus Pathogenesis

A Practical Approach

Edited by

MICHAEL B. A. OLDSTONE

*Department of Neuropharmacology,
Scripps Clinic & Research Foundation,
La Jolla, California, USA*

IRL PRESS
—at—
OXFORD UNIVERSITY PRESS
Oxford New York Tokyo

Oxford University Press
Walton Street, Oxford OX2 6DP

Oxford is a trade mark of Oxford University Press

Published in the United States
by Oxford University Press, New York

© *Oxford University Press, 1990*

All rights reserved. No part of this publication may be reproduced, stored in a retrieval system, or transmitted, in any form or by any means, electronic, mechanical, photocopying, recording, or otherwise, without the prior permission of Oxford University Press

This book is sold subject to the condition that it shall not, by way of trade or otherwise, be lent, re-sold, hired out, or otherwise circulated without the publisher's prior consent in any form of binding or cover other than that in which it is published and without a similar condition including this condition being imposed on the subsequent purchaser

British Library Cataloguing in Publication Data
Animal virus pathogenesis.
1. Animals. Bacterial diseases. Aetiology
I. Oldstone, Michael B. A. II. Series 576.6484
ISBN 0–19–963100–X
ISBN 0–19–963101–8 (pbk.)

Library of Congress Cataloging in Publication Data
Animal virus pathogenesis: a practical approach/edited by Michael B. A. Oldstone.
Includes bibliographical references.
(Practical approach series)
1. Virus diseases—Pathogenesis. I. Oldstone, Michael B. A. II. Series
QR201.V55A55 1990   616.'0194—dc20   89–71036
ISBN 0–19–963100–X
ISBN 0–19–963101–8 (pbk.)

Typeset by Cotswold Typesetting Ltd, Gloucester
Printed by Information Press Ltd, Oxford, England

Contents

List of Contributors	xiv
Abbreviations	xvi
Acknowledgements	xviii

1. Concepts of molecular pathogenesis   1
Michael B. A. Oldstone

1. Introduction   1
    Injury induced directly by viruses   1
    Injury induced indirectly by viruses   2
    Injury modulated by host genes   2
    Summary   4

  References   4

2. Molecular characterization of viral infections *in vivo*   5

2A. NUCLEIC ACID AND PROTEIN DETECTED MORPHOLOGICALLY WITH WHOLE ANIMAL BODY SECTIONS   5
W. Ian Lipkin and Michael B. A. Oldstone

1. Introduction   5

2. Sensitivity of WAS *in situ* hybridization and immunochemistry   5

3. Materials and methods   7
    Preparation of animals   7
    Cryomicrotomy   8
    In situ hybridization   9
    Hybridization   10
    Protein blotting   11

  References   13

Contents

2B. DETECTION OF MACROMOLECULES IMMOBILIZED ON SOLID SUPPORTS   14
Šárka O. Southern and Peter J. Southern

4. Introduction—antigens, antibodies, and protein gel techniques   14
5. Preparation and storage of protein samples   15
6. Preparation of antibodies   19
   Monoclonal antibodies   21
   Polyclonal antibodies   22
7. Labelling of purified antibodies and protein A with ^{125}I   23
8. Gel electrophoresis and electrophoretic transfer   27
   Preparation of antigen samples for electrophoresis   27
   Gel electrophoresis   30
   Electrophoretic transfer   31
9. Immunochemical probing of Western blots and autoradiography   34
   Treatment of blots before probing   35
   Probing of blots with antibodies and protein A   36
   Re-used probes, reprobed blots, and multiple antibody probing   38
10. RNA transfer and hybridization procedures   40
    Extraction of total cell RNA by treatment with guanidinium thiocyanate (GTC)   40
    Phenol extraction   43
    Minigel analysis of RNA samples   44
11. Oligo (dT) cellulose chromatography   45
12. RNA gel electrophoresis   45
13. Preparation of radioactively labelled hybridization probes   47
    Synthesis of double-stranded DNA probes   47
    Synthesis of single-stranded probes   49
    Gel transfer procedures   49
    Hybridization and washing conditions   50
    Reuse of hybridization filters   51
14. Recent technical developments   51
    Vacuum blotting   52
    Non-isotopic hybridization probe strategies   52
15. Ribonuclease contamination   53

    References   54

2C. NUCLEIC ACID EXTRACTION AND DETECTION FROM PARAFFIN-EMBEDDED TISSUES   55
Matthias Löhr and Michael I. Nerenberg

16. Introduction   55

17. General considerations   56

18. Sectioning paraffin blocks   57

19. Extraction of RNA and DNA   57

20. Analysis of nucleic acids   60

References   61

2D. DNA AMPLIFICATION/POLYMERASE CHAIN REACTION   62
Matthias Löhr and Maria S. Salvato

21. Introduction   62

22. PCR selection and preparation of primers   62

23. Preparation of DNA   63

24. DNA amplification   63
    PCR general considerations   64
    PCR   64

25. Cloning and sequencing of PCR products   65

26. RNA as a template for PCR   65

References   66

2E. TECHNIQUES FOR DOUBLE-LABELLING VIRUS INFECTED CELLS   67
Catherine Reynolds-Kohler and Jay A. Nelson

27. Introduction   67

28. Preparation of slides and specimens   68

29. Immunocytochemistry   71
    Detection systems   71
    Staining   73
    Immunocytochemistry: a method for single-labelling   75
    Immunocytochemistry: a method for double-labelling   77

Contents

30.	*In situ* hybridization method	78
31.	Immunocytochemistry—*in situ* double-labelling technique	84
	References	85

3. Detection of immune complexes and antibody mediated target cell lysis 87
Michael B. A. Oldstone

1. Immune complexes in body fluids during viral infection   87
    Test to detect immune complexes   87
    Purification of Clq   88
    Radiolabelling of Clq   89
    Standard curve and interpretation   90

2. Infectious virus–antibody immune complexes in body fluid   91
    Detection of infectious virus–antibody immune complexes   91

3. Tissue deposited virus–antibody immune complexes   92
    Detection of tissue bound virus–antibody immune complexes   92

4. Antibody mediated injury of virus infected cells   92
    Antibody and complement mediated injury   92
    Antibody-dependent cell mediated cytotoxicity   93

4. Detection, generation, and use of cytotoxic T lymphocyte (CTL) clones 95

4A. DETECTION OF CTL ACTIVITY   95
Michael B. A. Oldstone

1. Introduction   95
    Selection of target cells   95
    Preparation of lymphocytes from blood   96
    ^{51}Chromium-labelling of cells   97
    Effector CTL   98
    Evaluation and calculation of CTL assay   98

4B. GENERATION AND USE OF CTL CLONES   99
Hanna Lewicki

2. Primary CTL response   99

Contents

3. CTL clones — 99

4. Harvesting and use of peritoneal macrophages — 101

5. Use of CTL clones *in vivo* — 104
Adoptive transfer of cloned CTL — 104

6. Ablation of the host's immune response — 104

4C. USE *IN VITRO* TO MAP CTL EPITOPES — 104
J. Lindsay Whitton and Antoinette Tishon

7. Introduction — 104

8. Identification of virus target protein — 105
Use of reassortant viruses — 105
Use of recombinant DNA techniques — 105

9. Mapping viral epitopes seen by CTL — 113

4D. USE *IN VIVO* — 115
Linda S. Klavinskis and Michael B. A. Oldstone

10. Introduction — 115

11. Preparation of CTL clones for *in vivo* studies — 116
General considerations — 116

12. Removal of dead cells — 116

13. *In vivo* procedures: immunosuppression — 118

14. *In vivo* inoculation — 118

15. Migration studies *in vivo* — 119

16. Measurement of the ability of cloned CTL, as compared with freshly isolated lymphocytes, to migrate to lymphoid tissue — 119

References — 120

5. Methods for studying mouse natural killer cells — 121
Raymond M. Welsh

1. Introduction — 121

2. Histochemical staining — 122
Reagents — 122
Interpretation — 122

3.	Cytotoxicity assays	123
	Chromium-release assay	123
	Choice of target cells	124
	Single cell cytotoxicity assay	125
	Selection of target cell	125
	Modification of the single cell assay to quantitate lysis mediated by blast cells	126
4.	Isolation and preparation of effector cells	126
	Purification of NK cells	128
	Purification of endogeneous spleen NK cells	129
	Purification of activated spleen cells	129
	Purification of activated peritoneal NK cells	130
	Purification of NK cells by flow cytometry	130
5.	Modulation of NK cell activity and number *in vivo*	130
	Depletion of NK cell activity *in vivo*	130
6.	Demonstrating the anti-viral roles for NK cells *in vivo*	131
	Progression of infection under conditions of NK cell depletion	131
	Progression of virus infection in NK cell-deficient mice	132
	Adoptive transfers of NK cells into NK cell-deficient mice	132
7.	Generation of lymphokine activated killer cells	132
8.	Analyses of steps in the NK cell lytic cycle	132
	Binding	133
	Calcium flux	133
	Reorientation of golgi apparatus	133
	Secretion	133
	Lysis	134
9.	Measurement of cell migration	134
	Acknowledgements	134
	References	135

6. Viral–receptor binding assays 137

Peter L. Schwimmbeck, Matthias Löhr, and Antoinette Tishon

1.	Introduction	137
2.	Binding to immobilized receptor	137
	Permissive cells	137
	Preparation of membrane-enriched fractions	138
	SDS-gel	140
	Transfer to nitrocellulose	142
	Virus binding assay	143

Contents

3. Binding to receptor on cells — 144
   Purification of virus — 144
   Biotinylation of virus — 145
   Binding of virus — 146

References — 147

7. Viral-autoimmune experimental models — 149
Robert S. Fujinami

1. Introduction — 149
2. Basic techniques and their utilization — 150
   Identification of common regions — 150
   Determination of sites for disease induction — 150
   Peptides — 151
   Adjuvant — 151
   Injection scheme — 152
   Determination of antibody production — 152
   Determination of cellular reactivity — 158
3. Determination of injury or pathology — 161

Acknowledgements — 161

References — 161

8. Synthetic peptides as antigens — 163
Thomas Dyrberg and Hans Kofod

1. Introduction — 163
2. Selection of sequences — 164
3. Peptide synthesis — 165
4. Purification of peptides — 165
5. Peptide conjugation to carrier protein — 166
6. Immunization — 169

Acknowledgements — 170

References — 170

Index — 173

Contributors

THOMAS DYRBERG
Hagedorn Research Laboratory, Niels Steensensvej 6, Dk-2820 Gentofte, Denmark.

ROBERT FUJINAMI
Department of Pathology, University of California at San Diego, La Jolla, CA 92093, USA.

LINDA KLAVINSKIS
Roche Products Limited, Department of Chemotherapy Biology, P.O. Box 8, Welwyn Garden City, Hertfordshire AL7 3AU, UK.

HANS KOFOD
Hagedorn Research Laboatory, Niels Steensensvej 6, Dk-2820 Gentofte, Denmark.

HANNA LEWICKI
Department of Neuropharmacology (IMM-6), Scripps Clinic & Research Foundation, 10666 N. Torrey Pines Road, La Jolla, CA 92037, USA

W. IAN LIPKIN
Department of Neurology, University of California at Irvine, Irvine, CA 92717, USA.

J. MATTHIAS LÖHR
Department of Internal Medicine I, University of Erlangen, Krankenhaus Str. 12, D-8520 Erlangen, Germany.

JAY NELSON
Department of Immunology (IMM-1), Scripps Clinic & Research Foundation, 10666 N. Torrey Pines Road, La Jolla, CA 92037, USA.

MICHAEL I. NERENBERG
Department of Neuropharmacology (IMM-6), Scripps Clinic & Research Foundation, 10666 N. Torrey Pines Road, La Jolla, CA 92037, USA.

MICHAEL B. A. OLDSTONE
Department of Neuropharmacology (IMM-6), Scripps Clinic & Research Foundation, 10666 N. Torrey Pines Road, La Jolla, CA 92037, USA.

C. REYNOLDS-KOHLER
Department of Immunology (IMM-1), Scripps Clinic & Research Foundation, 10666 N. Torrey Pines Road, La Jolla, CA 92037, USA.

Contributors

MARIA S. SALVATO
Department of Neuropharmacology (IMM-6), Scripps Clinic & Research Foundation, 10666 N. Torrey Pines Road, La Jolla, CA 92037, USA.

PETER SCHWIMMBECK
Department of Internal Medicine and Cardiology, University of Dusseldorf, Moorenstrasse 5, D-4000 Dusseldorf, Germany.

PETER J. SOUTHERN and ŠÁRKA O. SOUTHERN
University of Minnesota, Department of Microbiology, 1460 Mayo Bldg., Box 196 UMHC, 420 Delaware St., S.E. Minneapolis, MN 55455, USA.

ANTOINETTE TISHON
Department of Neuropharmacology (IMM-6), Scripps Clinic & Research Foundation, 10666 N. Torrey Pines Road, La Jolla, CA 92037, USA.

RAYMOND WELSH
Division of Pathology, University of Massachusetts Medical Center, 55 Lake Avenue North, Worcester, MA 01605, USA.

J. LINDSAY WHITTON
Department of Neuropharmacology (IMM-6), Scripps Clinic & Research Foundation, 10666 N. Torrey Pines Road, La Jolla, CA 92037, USA.

Abbreviations

AAP	avidin–alkaline phosphatase
AB	Amido black
ABC	avidin–biotin complex
ADCC	antibody-dependent cell-mediated cytotoxicity
AEC	3 amino-9 ethylcarbazole
APC	antigen presenting cells
APS	ammonium persulphate
BSA	bovine serum albumin
BUdR	bromodeoxy uridine
CF	cytotoxic factor
CMC	carboxymethyl cellulose
CNS	central nervous system
CTL	cytotoxic T lymphocytes
DAB	3,3′ diaminobenzadine
ddH_2O	deionized distilled water
EAE	experimental allergic encephalomyelitis
ELISA	enzyme linked immunoadsorbent assay
EMCV	endomyocarditis virus
FACS	fluorescence activated cell sorter
FCS	fetal calf serum
FITC	fluorescein isothiocyanate
GAG	glucose oxidase–anti-glucose oxidase
GTC	guanidinium thiocyanate *or* (cf. Ch. 2.3, p. 99) guanidium isothiocyanate
LAK	lymphokine activated killer
LCMV	lymphocytic choriomeningitis virus
LGL	large granular lymphocytes
MPB	myelin basic protein
MEM	modified Eagles medium
MHC	major histocompatibility complex
MHV	mouse hepatitis virus
NK	natural killer
PAGE	polyacrylamide gel electrophoresis
PAP	peroxidase–anti-peroxidase
PBL	peripheral blood lymphocytes
PBS	phosphate-buffered saline
PCR	polymerase chain reaction
PEG	polyethylene glycol
PLP	proteolipid protein
PLPG	paraformaldehyde-lysine-periodate-glutaraldehyde
SPA	staphylococcal protein A

Abbreviations

TCGF	T cell growth factor
TCR	T cell receptors
TK	thymidine kinase
TNF	tumour necrosis factor
VBS	veronal-buffered saline
VV	vaccinia virus
WAS	whole animal section

Acknowledgements

Many of the strategies to study viral pathogenesis recorded in this book were developed or extended by colleagues who have worked in the Viral-Immunobiology Unit at Scripps. Thanks to Raymond Welsh, Professor of Microbiology and Pathology at the University of Massachusetts Medical School, Worcester; Peter Southern, Associate Professor of Microbiology at the University of Minnesota Medical School, Minneapolis; Robert Fujinami, Association Professor of Pathology at the University of California at San Diego Medical School, La Jolla; Peter Schwimmbeck, Assistant Professor of Medicine at the University of Dusseldorf, Germany; Matthias Löhr, Assistant Professor of Medicine at the University of Erlangen, Germany; Ian Lipkin, Assistant Professor of Neurology at the University of California at Irvine; and Linda Klavinskis, Research Scientist at Hoffman-La Roche, London. Drs Whitton, Nerenberg, Salvato, and Nelson are currently on the faculty and Ms Tishon, Lewicki, and Reynolds-Kohler are currently on the technical staff at the Research Institute of Scripps. S. Southern is in Minneapolis, Minnesota. NIH support from the Allergy and Infectious Disease Institute, Neurologic Institute and Aging Institute was responsible, in large part, for funding the development and utilization of the techniques reported here. We thank Mrs Gay Schilling for typing and management of the book.

1

Concepts of molecular pathogenesis

MICHAEL B. A. OLDSTONE

1. Introduction

Pathology is the study of disease and pathogenesis is the study of the mechanism(s) by which disease occurs.

1.1 Injury induced directly by viruses

In general viruses cause disease by two distinct ways. First, according to their styles of replication and viral products made or host factors induced, viruses directly damage a cell. Hence, viruses directly destroy a cell by shutting down its protein synthesis, by manufacturing products toxic to the cell or by perturbing membranes either internal (i.e. lysosomal, endosomal) or external (i.e. plasma). Disturbance of internal membranes can result in release of a stable of enzymes potentially autolytic and thus dangerous to the cell. Disturbance of the plasma membrane can lead to cell–cell fusion, formation of giant syncytia and cell lysis or improper passage of ions and fluids causing cell distortion and death. Viruses may also affect cells directly by altering a cell's function without lysing the cell. This more subtle effect causes disease but does not cause cell lysis. According to this scenario the virus infects differentiated cells (i.e. neurons, lymphocytes, endocrine cells, etc.), and alters their synthesis of neurotransmitters, lymphokines, immunoglobulins, hormones, etc. In contrast to the other mechanisms mentioned previously whereby the end result is destruction of the infected cell, in this instance the architecture and morphology of the infected cell appears normal by light and high resolution microscopy or by assay of cell growth efficiency, total cell RNA, DNA or protein synthesis. Because less than the usual amount of the differentiated product is made, i.e. insulin, growth hormone, ablation of specific anti-viral cytotoxic T lymphocyte function and/or immunoglobulin production, homeostasis can be imbalanced and disease produced (diabetes, growth retardation, viral persistence). Examples of these two different pathways are easily catalogued with two different models of virus induced diabetes; i.e. diabetes caused by endomyocarditis virus (EMCV) or lymphocytic choriomeningitis virus (LCMV). Both viruses are tropic for and replicate in the β cells of the islets of Langerhans, the cells that synthesize insulin. In the first instance, the diabetic variant of EMCV lyses β cells, resulting in a dramatic fall in insulin secretion, resultant low plasma levels of insulin causing increased elevations of blood glucose. The resultant diabetes is severe and causes a metabolic death in infected mice. By contrast LCMV, while also replicating in β cells of the islets of Langerhans, does not destroy these cells. As a result of infection a mild decrease in insulin level ensues with a significant but mild elevation of blood glucose and abnormal glucose tolerance tests. Thus cell lysis or inflammatory response need not be a marker of viral induced disease.

Regardless of whether the virus is lytic or non-lytic, its fingerprints are represented in the targeted cell or tissues. *Figure 1a* shows that depending on what phase of viral replication is blocked, only a particular component of the virus is available for identification. If the replication is completed the biological activity (plaquing ability) of the virus can be determined. Procedures for detection of genome, message or protein are covered in Chapter 2.

1.2 Injury induced indirectly by viruses

Proteins encoded by viral genes represent potential antigens capable of eliciting specific immune responses. Such anti-viral immune responses can interact with virus or viral antigens in fluids or with viral antigens expressed on cell surface. By this means the host's immune response acts to clear virus from spreading and destroys the factories (cells) making progeny virus. The injurious outcome while usually beneficial to the host nevertheless is the cause of many of the clinical symptoms of viral infections (rash, myalgia, central nervous system disorders, etc.). Further such immune mediated (immunopathological) injury when occurring in the meninges of the brain, the brain cerebrum or heart, etc., can produce disastrous results. A flow chart of immunopathology mediated injury is given in *Figure 1b* and procedures to detect or study these components of the immune system are detailed in Chapters 3, 4 and 5.

1.3 Injury modulated by host genes

A detailed description of the role played by host genes in viral pathogenesis is outside the scope of this volume. Interested readers are referred to the appropriate reference at the end of this chapter. However, two points need to be emphasized. First, in an experimental host with well defined genetics, genetic linkage maps can usually be obtained. For example, when a single dominant or recessive gene is determined by Mendellian genetics, recombinant inbred mice can be used for fine dissection and location of gene encoding susceptibility or resistance. Similar use of somatic cell fusion with chromosome deletions provides a powerful analytic complementary approach for *in vitro* analysis. Second, immune responses, both humoral (antibody) and cell mediated (cytotoxic T lymphocyte) are under the control of immune response genes that have been best mapped in mice and man. Experimentation indicates a hierarchy for immune response genes to occur.

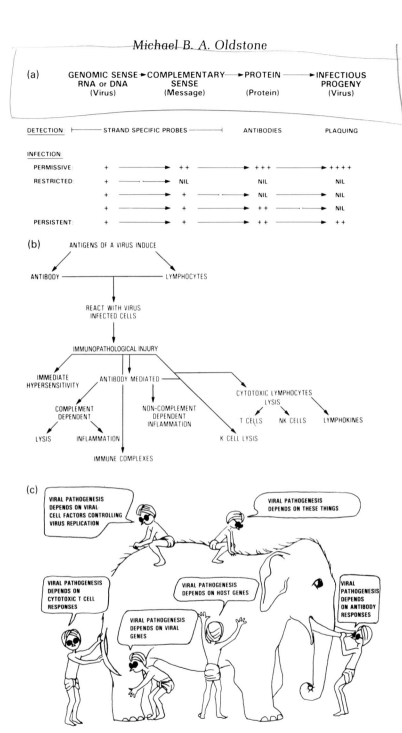

Figure 1(a). Replicative strategy followed by viruses and the probes required to detect virus during various stages of its replication. **(b)** Mechanism(s) of how virus induces immune response disease (immunopathology). **(c)** Definition of viral pathogenesis.

1.4. Summary

Understanding viral pathogenesis requires knowledge and use of techniques in several disciplines (*Figure 1c*). Critically addressing research in this area requires working at the interface of virology and immunology.

References

1. Wagner, R. R. (1984). In *Comprehensive Virology*, (eds. H. F. Conrat and R. W. Wagner), Vol. 19, pp. 1–64. Plenum Press, New York.
2. Oldstone, M. B. A. (1989). *J. Infect. Dis.*, **159**, 384.
3. Whitton, J. L. and Oldstone, M. B. A. (1990). In *Virology*, (ed. B. N. Fields), Raven Press Ltd., New York, pp. 369–381.
4. Oldstone, M. B. A. and Lampert, P. W. (1979). Springer Seminars in Immunopathology.
5. Oldstone, M. B. A. (1975). In *Progress in Medical Virology*, (ed. J. L. Melnick), Vol. 19, p. 84. S. Karger, Basel.

2

Molecular characterization of viral infections *in vivo*

2A. NUCLEIC ACID AND PROTEIN DETECTED MORPHOLOGICALLY WITH WHOLE ANIMAL BODY SECTIONS

W. IAN LIPKIN and MICHAEL B. A. OLDSTONE

1. Introduction

Natural and experimental infections of animals continue to be important as models in studying the pathogenesis and management of infectious diseases. Classically, methods for examining animal models have required extraction of individual tissues for biochemical or histological analysis. Examples of these techniques reviewed elsewhere in this volume include nucleic acid and protein electrophoresis, immunohistochemistry and enzyme linked immunoadsorbent assays (ELISA). Over the past few years, however, we have used a new technology which allows for sensitive and specific detection of nucleic acids and proteins in whole animal sections (WAS) fixed to tape or artificial membranes. The power of WAS technology rests in its unique potential for defining and quantitating the distribution of genes and gene products in an anatomical context. Further, because the entire animal is examined, there is no bias *a priori* against recognizing novel distributions of genes or gene products. Although we have used WAS primarily for studies in viral pathogenesis, problems in developmental biology, neoplasia, and pharmacology may also be suitable for WAS analysis. WAS would also seem ideal for the investigation of tissue-specific distributions of transcripts and proteins in transgenic animal systems. In this chapter we will discuss our methods for WAS *in situ* hybridization and protein blotting (*Figure 1*). For illustrations of WAS applications the reader is referred to a recent review (1).

2. Sensitivity of WAS *in situ* hybridization and immunochemistry

The sensitivity of WAS *in situ* hybridization and protein blots for detection of viral sequences and antigens has been well studied in the murine model of

Using whole animal sections

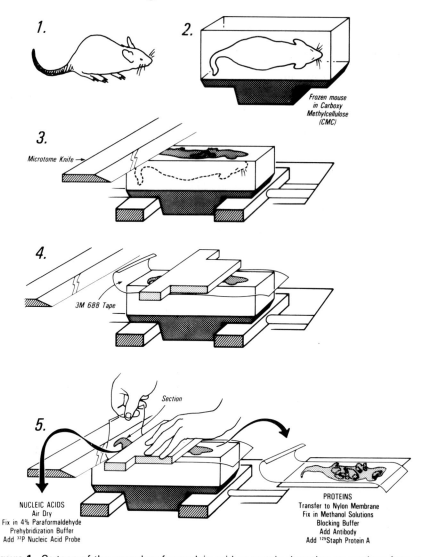

Figure 1. Cartoon of the procedure for nucleic acid or protein detection on sections from a whole animal. Procedure has been used successfully to study mice, rats, guinea pigs, hamsters, woodchucks, rabbits, and monkeys.

persistent infection with lymphocytic choriomeningitis virus (LCMV). In the LCMV model, WAS *in situ* hybridization is 10% as sensitive as slot blot hybridization for detection of LCMV RNA in infected mouse tissues. The threshold for detection of LCMV antigens in WAS protein blots can be as low as 10–20 ng of protein/cm^2.

3. Materials and methods

Materials and equipment required are listed as follows:

For WAS Cryomicrotomy
- cryomicrotome #2258 or #2250 (Pharmacia LKB, Gaithersberg, MD)
- freezing frames (Pharmacia LKB, Gaithersberg, MD)
- 3.5% carboxymethylcellulose (Sigma, St Louis, MO)
- dry ice
- ethanol
- #688 transparent adhesive tape (Pharmacia LKB, Gaithersberg, MD)
- electric razor

For WAS in situ hybridization
- 4% buffered paraformaldehyde (pH 7.4)
- phosphate-buffered saline
- deionized formamide
- 20 × SSC
- Denhardt's solution (2)
- temperature-controlled water baths or incubators (37–55 °C)
- haematoxylin
- eosin
- sealable plastic bags and sealing apparatus
- XAR-5 Film (Eastman Kodak Co., Rochester, NY)
- cronex lightning-plus intensifying screens (E.I. DuPont de Nemours & Co., Inc., Wilmington, DE)
- ^{32}P-labelled cDNA or RNA probes
- saran wrap (Dow, Indianapolis, IN)
- film cassettes

For WAS protein blots
- 0.2 μm Biotrans membranes (ICN Biomedicals, Inc., Costa Mesa, CA)
- Laemmli buffer (3)
- methanol
- Blotto (4)
- lithium chloride wash (0.5 M lithium chloride, 0.1 M Tris, pH 8.0, 1% NP-40)
- primary antibodies
- secondary antibodies
- [^{125}I]staphylococcal protein A
- XRP-1 Film (Eastman Kodak Co., Rochester, NY)
- Film cassettes

3.1 Preparation of animals

Anaesthetize animals with an appropriate inhalant or parenteral agent and kill them through exsanguination. The latter may be accomplished either through phlebotomy or cardiac perfusion. Exsanguination decreases background noise in protein blots, presumably, because it reduced binding of staphylococcal protein A to immunoglobulins and lymphocytes in the vascular system. Phlebotomy is readily achieved through deep axillary incision and has the advantage of maintaining thoracic and abdominal integrity. When analysing models in which viral pathogens are found in high titre in blood or serum, however, cardiac perfusion may be preferred. We typically perfuse with phosphate-buffered saline (PBS). Perfusion with fixative solutions such as paraformaldehyde can interfere with transfer of nucleic acids and proteins from tissue sections to artificial membranes.

After killing, the animals should be shaved with an electric razor in preparation for freezing in mounting media. This is necessary to prevent fur from becoming interposed between the section and the tape or, between the section on tape and the membrane to which the section will be transferred. The animal is then frozen into a block of 3.5% carboxymethyl cellulose (CMC) on top of a base suitable for use in a cryomicrotome (*Figure 1*). It is critical that the animals be mounted in the block such that sections will represent regions of interest. We have found the right lateral decubitus position the most useful. This orientation allows one to simultaneously examine brain, spinal cord, salivary gland, thymus, heart, lungs, liver, spleen, kidney, gastrointesttinal tract and genitourinary tract. Place the base on to which the animal will be frozen within a freezing frame, add CMC and position the animal as described above. Then lower the base into a bath of dry ice–ethanol. The ethanol should not come into contact with the CMC. If it does, the CMC will freeze only to a slush consistency, rendering sectioning difficult. Once the CMC freezes, remove the block on its base from the freezing frame and place it in aluminium foil at $-20\,°C$. Under these conditions, animals can be stored for at least one month without appreciable loss of protein signal. The shelf life is even longer (6 months) for *in situ* hybridization studies.

3.2 Cryomicrotomy

Remove the block to be sectioned from $-20\,°C$ to the cryomicrotome cabinet and allow 2–4 h to equilibrate to the temperature within the cabinet (-15 to $-20\,°C$). An old knife may be used to trim the block until the plane of interest is reached. Then switch to a sharper blade to cut specimens for analysis. This ensures sections of uniform thickness without nicks. Collect 20- or 40-μm sections on transparent adhesive tape (#688 3 M). Either transfer these sections immediately to membranes for *in situ* hybridization or protein blotting or allow them to thaw and dry for 10–30 min at room temperature for *in situ* hybridization. Some workers allow sections to dry within the cryomicrotome chamber at -15 to $-20\,°C$. In our hands this approach has led to cracking and detaching of sections from tape. For reasons unclear to us, different lots of #688 tape vary in their capacity to retain tissue during drying, *in situ* hybridization or post hybridization washes. We recommend pilot experiments with different lots of tape, before purchasing large stocks. Tape can be stored for several years in airtight containers away from heat and moisture.

3.3 *In situ* hybridization

See *Protocol 1* and *Figure 2*. In situ hybridization experiments can be performed either with sections on tape or, with membranes to which nucleic acids have been transferred. We prefer to use secions on tape for two reasons: first, a haematoxylin and eosin stained sections show organ anatomy with remarkable clarity; second, it eliminates the possibility of incomplete nucleic acid transfer. After sections on tape have air-dried, they are immersed in freshly prepared 4% paraformaldehyde pH 7.4 for 30 min. We have used other fixatives including paraformaldehyde-lysine-periodate (PLP) (5) and paraformaldehyde-lysine-periodate-glutaraldehyde (PLPG) (6) however, these have not improved tissue morphology or the quality of *in situ* hybridization autoradiographs. In addition, sections fixed with PLP have detached from tape during post hybridization washes. Following fixation, sections are washed in PBS then either used immediately for *in situ* hybridization or stored, tissue side down against clean

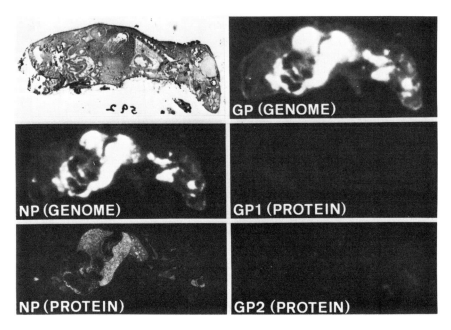

Figure 2. Detection of genes and proteins expressed in tissues of an adult mouse infected since birth with lymphocytic choriomeningitis virus. Thirty micron consecutive sections were hybridized to a ^{32}P-labelled cDNA probe to the viral nucleoprotein (NP) or viral glycoprotein (GP). To detect protein, monoclonal antibodies to NP or GP (glycoprotein is post-translationally cleaved to GP1 and GP2 protein) were used. Restricted expression of GPs compared to NP is shown. The panel in the **upper left** shows a haematoxylin–eosin stain of the 30-μm section.

glass plates, at 4 °C. We have successfully used year-old sections stored in this way.

Protocol 1. WAS *in situ* hybridization

1. Collect tissue sections on tape and allow them to thaw and dry for 15–30 min.
2. Fix sections for 30 min in buffered 4% paraformaldehyde (pH 7.4).
3. Wash sections for 5 min × 3 in phosphate-buffered saline (total 15 min).
4. Prehybridize sections in sealed plastic bags in 50% deionized formamide, 5 × SSC, 2.5 × Denhardt's, 150 µg/ml sonicated salmon sperm DNA for 4–12 h at 37 °C (cDNA probes) or 55 °C (RNA probes). Use 0.5–1.0 ml of solution/10 cm^2 of tissue section surface area.
5. Replace prehybridization solution with half the volume of fresh prehybridization solution containing ^{32}P-labelled cDNA or RNA probe in a concentration of $1-2 \times 10^6$ c.p.m./ml. Hybridize at either 37 °C (cDNA probes) or 55 °C (RNA probes) for 24–48 h.
6. Remove hybridization solution.
7. Wash sections in 2 × SSC at 37 °C for 30 min, in 2 × SSC at 55 °C for 30 min then in 0.1 × SSC at 55 °C for 30 min. **DO NOT** use SDS (sections will detach from tape).
8. Stain sections in haematoxylin and eosin.
9. Set sections against saran wrap. Cover sections with saran wrap and expose to XAR-5 film with an intensifying screen at −70 °C.

3.4 Hybridization

Hybridization conditions will depend on the nature of the probes employed. We have used both DNA and RNA probes. RNA probes may be more sensitive and have the advantage of allowing one to distinguish between transcripts of opposite polarity. The latter has been an important advantage in studying arenaviruses like LCMV which have an ambisense replication strategy. RNA probes do, however, have a higher hybrid melting temperature than equivalent cDNA probes (7) necessitating use of high stringency conditions for hybridization and post hybridization washes as well as RNase treatments. We label probes with ^{32}P because its higher specific activity allows for short exposures with good regional localization. cDNA probes are labelled via either nick translation (8) or random hexanucleotide primer (9, 10) reactions to a specific activity of $1-5 \times 10^8$ c.p.m./µg of DNA. Use of purified DNA inserts, rather than whole plasmid DNA, is helpful in optimizing background. RNA probes are transcribed using protocols recommended by Promega Biotech, the manufacturer of the vectors we use most commonly.

Sections are prehybridized for 4 h–overnight in 50% formamide, 5 × SSC,

2.5 × Denhardt's solution (2), 150 μg/ml boiled, sonicated salmon sperm DNA. They are then hybridized for 24–48 h in freshly prepared prehybridization solution to which probe has been added to a concentration of 1–2 × 10^6 c.p.m./ml. Prehybridization and hybridization are performed in plastic bags (seal-a-meal) at either 37 °C (cDNA hybridization) or 55 °C (RNA hybridization) in a volume of 0.5–1 ml of solution/10 cm^2 of tissue surface area. After cDNA hybridization, sections are washed first in 2 × SSC at 37 °C for 30 min, then in 2 × SSC at 55 °C for 30 min. RNA hybridizations are followed by two 30-min washes in 0.1 × SSC at 55 °C and one wash in 0.1 × SSC at 55 °C for 30 min. Use of SDS, or other detergents, should be avoided because sections will detach from tape. Following the final wash, sections are stained with haematoxylin and eosin, set against saran wrap to dry and exposed to film (Kodak XAR-5) at −70 °C with an intensifying screen for 24–72 h. Stained tissue sections are used to identify organ anatomy in autoradiographs.

3.5 Protein blotting

See *Protocol 2* and *Figures 2* and *3*. Sections are placed on to 0.2 μm nylon membranes (ICN Biomedicals Inc., Costa Mesa, CA) to allow passive protein transfer for 10–30 min. The membranes with sections attached are immersed first in Laemmli/Buffer (3) and methanol then placed in acetic acid and methanol. This fixes proteins to the membranes and dissolves the tape adhesive. The tape is stripped from the membranes and the membranes are rinsed in water. Tissues adherent to the membranes is stripped away with a glass slide or a razor blade. Membranes are allowed to dry and either used immediately or stored at 4 °C. To facilitate identification of organ anatomy in protein blot autoradiographs we collect adjacent sections for staining with haematoxylin and eosin.

To block non-specific protein reactive sites, membranes are incubated in Blotto (4). We have used both polyclonal and monoclonal primary antibodies for detection of viral and host proteins. Membrane preparation can obscure target epitopes through protein denaturation. For this reason we recommend use of polyclonal antibodies. Polyclonal antibodies are often reactive with gastrointestinal flora. Adsorption of primary antibodies with a minced preparation of rodent intestines reduces this background reactivity. The optimal concentration of primary antibody varies but is usually 5–10-fold higher than that required for immunohistochemistry in cryostat sections. Membranes are incubated in primary antibody diluted in Blotto, rinsed in Blotto then incubated in [^{125}I]staphylococcal protein A (SPA). ^{125}I-SPA binds well to antibodies prepared in rabbits, and mice. If one is using an antibody prepared in a different species it may be necessary to use a secondary antibody which binds SPA. A listing of the relative affinity of SPA for immunoglobulins derived from different species can be found in Langone (1982) (11). After antibody incubation(s) membranes are washed first in Blotto then in 0.5 M LiCl, 0.10 M Tris, pH 8.0, 1% NP-40. They are rinsed in water, drained and set directly against a fine grain film (XRP-1) Kodak) at room temperature for 24–72 h.

Using whole animal sections

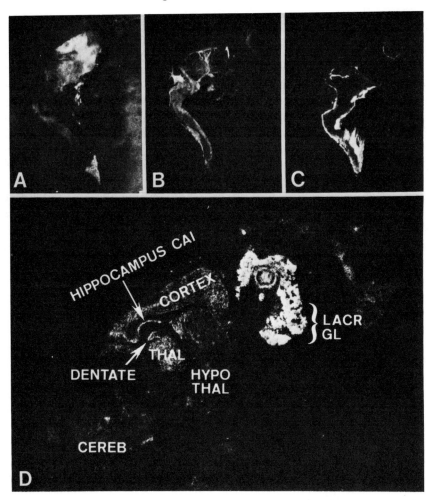

Figure 3. Detection of proteins in a whole mouse animal section, focus on central nervous system (CNS). **Panel A** uses a monoclonal antibody to Theta 1.2 to show antigen distribution in the brain and thymus; **panel B** uses an antibody to myelin basic protein to show its distribution in the CNS; **panel C** employs an antibody to glial fibrillary acid protein; **panel D** records a 10 × enlargement of a brain section from a 3–4-month-old mouse, persistently infected with LCMV. The viral nucleoproteins can be traced in several CNS neuronal tracts (CEREB, cerebellum; THAL, thalamus; HYPOTHAL, hypothalamus; LACR GL, lacrimal gland). This anatomical distribution of viral protein was observed using a rabbit antibody to LCMV nucleoprotein followed by ^{125}I[staph A]. Recording of viral localization has been done successfully in this laboratory for wild mouse ecotropic virus, reovirus, herpes simplex virus, rabies virus, polyoma virus in mice, woodchuck hepatitis virus in woodchucks.

Protocol 2. WAS protein blotting

1. Place sections on tape against 0.2 µm nylon membranes. Allow 30 min for protein transfer.
2. Float sections and attached membranes in 70% Laemmli buffer and 30% methanol for 10 min.
3. Float sections and attached membranes in 10% acetic acid and 10% methanol for 30 min. Tape will begin to detach from membranes.
4. Remove tape and residual tissue from membranes. Rinse membranes in distilled water for 10 min.
5. Incubate membranes in Blotto (5% non-fat dry milk, 0.01% anti-foam A and 0.001% thimerosal in PBS) for 2–4 h at room temperature or, for 12–18 h at 4 °C.[a]
6. Incubate membranes in primary antibody diluted in Blotto for 4 h at room temperature or, for 12–18 h at 4 °C.
7. Wash membranes in Blotto for 15 min × 3 (total 45 min).
8. Incubate membranes with [^{125}I]staphylococcal protein A in Blotto for 1 h. Use $0.5–1.0 \times 10^6$ c.p.m. of [^{125}I]staphylococcal protein A/ml of Blotto.
9. Remove [^{125}I]staphylococcal protein A/Blotto solution.
10. Wash membranes in Blotto for 10 min × 3 (total 30 min).
11. Wash membranes in 0.5 M lithium chloride, 0.5 M Tris, pH 8.0, 1% NP-40 for 30 min.
12. Rinse membranes in distilled water for 10 min.
13. Allow membranes to dry.
14. Expose membranes against XRP-1 film at room temperature for 24–72 h.

[a] Incubations can be performed in either sealed plastic bags or in dishes. If dishes are used membranes should be free to float in solutions. If plastic bags are selected allow 0.5–1.0 ml of solution/10 cm of membrane surface area.

References

1. Lipkin, W. I., Villarreal, L. P., and Oldstone, M. B. A. (1988). *Curr. Topics Microbiol. Immunol.*, in press.
2. Maniatis, T., Fritsch, E. F., and Sambrook, J. (1982). *Molecular Cloning.* Cold Spring Harbor Laboratory, New York.
3. Laemmli, U. K. (1970). *Nature*, **227**, 680.
4. Johnson, D. A., Gautsch, J. W., Sportsman, J. R., and Elder, J. H. (1984). *Gene. Anal. Tech.* **1**, 3.
5. McLean, I. W. and Nakane, P. K. (1974), *J. Histochem. Cytochem.* **22**, 1077.
6. Gendelman, H. E., Moench, T. R., Narayan, O., and Griffin, D. E. (1983). *J. Immunol. Methods*, **65**, 137.

7. Cox, K. H., DeLeon, D. V., Angerer, L. M., and Angerer, R. C. (1984). *Dev. Biol.*, **101**, 485.
8. Rigby, P. W. J., Dieckmann, M., Rhodes, C., and Berg, P. (1977). *J. Mol. Biol.*, **113**, 237.
9. Feinberg, A. and Vogelstein, B. (1983). *Anal. Biochem.*, **132**, 6.
10. Feinberg, A. and Vogelstein, B. (1984). *Anal. Biochem.*, **137**, 266.
11. Langone, J. J. (1982). *Adv. Immunol.*, **32**, 157.

2B. DETECTION OF MACROMOLECULES IMMOBILIZED ON SOLID SUPPORTS
ŠÁRKA O. SOUTHERN and PETER J. SOUTHERN

4. Introduction—antigens, antibodies, and protein gel techniques

Immunochemical protein blotting has become a widely employed method for the detection and biochemical characterization of proteins (12–15). Protein blotting, like the equivalent nucleic acid transfer and hybridization methods, permits sensitive analysis of a specific target in a sample containing a vast excess of unrelated macromolecules. Samples for protein blotting are usually crude, unseparated lysates of cells or microorganisms, or various biological fluids.

The basic immunoblotting procedure, also called Western blotting, involves SDS polyacrylamide gel electrophoresis (PAGE) separation and electrophoretic transfer (blotting) of proteins on to a solid support, usually a nitrocellulose or nylon filter, followed by reaction with an antibody, as detailed in *Protocol 3*.

Protocol 3. Basic Western blotting procedure

1. Prepare a (crude) sample of the protein.
2. Obtain or prepare a specific antibody which can recognize a denatured form of the protein antigen.
3. Separate the antigen sample by electrophoresis in SDS-PAGE gel. Prior to the electrophoresis, the antigen sample can be subjected to various treatments such as denaturation by SDS, reduction of S–S bonds, chemical cleavage or enzymatic modification.
4. Transfer the antigen from the SDS-PAGE gel on to a solid matrix using electroblotting.
5. React with specific antibody followed by a reporter molecule such as ^{125}I-labelled secondary antibody or protein A, or an enzyme–antibody conjugate.
6. Visualize the antigen band(s) by autoradiography or colorimetry.

Protocol 3. *Continued*

7. Assess specific properties of the visualized antigen(s): molecular weight, presence of S–S bonds or carbohydrate moieties.

The antigen–antibody complexes formed on the filter are then visualized through a radiographic or colorimetric technique. Polyclonal rabbit or murine antibodies raised against denatured proteins, and ^{125}I-labelled protein A have been among the most commonly used reagents so far but monoclonal antibodies can be also used successfully (*Table 1*).

In its basic form, protein blotting can provide information about the molecular weight and molecular complexity of a specific antigen, and the stability of the detected epitopes. However, protein blotting is a very versatile method allowing diverse variations on the basic procedure that can be tailored according to a particular experimental problem:

- Blotting from 2D gels, silica gel thin-layer plates or IEF gels.
- Analysis of antigens modified by chemical or enzymatic means such as reduction of S–S bonds, removal of glycosyl groups, or partial proteolysis.
- Analysis of non-covalently linked polypeptide complexes: discrimination between a complex of antigenically related molecules (e.g. a homodimer) and a complex involving the antigen molecule associated with unrelated proteins (e.g. a heterodimer). The complex is dissociated and resolved by SDS-PAGE electrophoresis, and blotted. Only molecules bearing the specific epitope are detected on Western blots.
- Dot immunobinding: protein samples are directly spotted on to filters and incubated with antibody (13, 16). This technique is useful for rapid assessment of antigen concentration, and epitope stability properties.
- Plaque blotting (virus or phage): analysis of bacteriophage expression libraries, or viral infections in tissues and cultured cells.
- Transfer from thin tissue sections and cell spreads. Mice and other small animals can be analysed on whole-body section blots (17).
- Reverse blotting: a specific antibody can be detected and quantitated by reaction with the antigen using the Western blotting or dotting formats. This approach is useful for analysis of serum samples or screening of hybridoma supernatants.
- Preparative blotting allows recovery of a transferred protein antigen, or direct immunization of animals with a filter-bound antigen (18).
- Ligands such as hormones (19, 20), enzyme substrates (21), whole cells (22), purified viral particles (23), or nucleic acid fragments (24, 25) can be used as probes for detection of specific proteins instead of antibodies.

5. Preparation and storage of protein samples

In order to analyse a protein by Western blotting it is necessary to solubilize the starting biological material and prepare a homogeneous protein extract. The

Table 1. Systematic assessment of antigen–antibody interactions in a dot blot format

Categories of antigen–antibody interaction	Pretreatment of protein extract		
	None[a]	SDS[b]	SDS, Reduced[c]
A	+	+	+
B	+	+	−
C	+	−	−
D	−	−	−

Interpretation of the different results
Category
A The antibody is able to recognize a denatured and reduced antigen, and can be therefore used in a standard Western blotting procedure (*Sections 8 and 9*) involving reducing SDS-PAGE electrophoresis.
B The antibody can recognize a denatured antigen and can be therefore used in a standard Western blotting procedure (*Sections 8 and 9*) involving SDS-PAGE electrophoresis under non-reducing conditions.
C The antibody is not able to recognize a denatured antigen but should be nevertheless tested for use in Western blotting under modified conditions: substitute CHAPSO for SDS in SDS-PAGE gel, renature the antigen in the gel or after transfer to a solid support (*Section 8.2*), and avoid heat-denaturation during electrophoresis and electroblotting.
D The antibody is not capable of detecting antigen under the present conditions and Western blotting should not be performed without further testing:
  1. Increase the antigen concentration by initial immunoprecipitation or affinity chromatography, and change from nitrocellulose to nylon for the blotting (*Sections 8.1 and 8.3*).
  2. Use an amplified detection system for antigen visualization (*Section 9*).
  3. Test other protein extraction procedures (*Section 5*).
  4. Investigate the properties of other antigen-specific antibodies. Usually, antibodies raised against denatured soluble immunogens are more likely to detect the denatured protein antigen following electrophoretical transfer from SDS-PAGE gels, than antibodies (mAbs in particular) raised against whole cells or native protein molecules (*Section 6*).

[a] Protein extracts prepared by cell lysis with a non-denaturing detergent such as Nonidet P-40, Triton X-100 or CHAPSO. Dot the antigen extract in three serial dilutions, e.g. $5 \times 10^6/10^6/2 \times 10^5$ cell equivalents, on to a nitrocellulose filter. Probe the dot blot with a specific mAb or a polyclonal antibody as outlined in *Section 9*.
[b] Add 20% SDS to the extract so that the final concentration is 0.5% and incubate at room temperature for 30 min before dotting.
[c] Add 20% SDS and 14.2 M (neat) 2-mercaptoethanol to the extract so that the final concentrations are 0.5% and 2%, respectively. Boil the mixture for 5 min and chill on ice before dotting.

protein sample can be a crude preparation such as a detergent lysate of tissues, whole cultured cells, blood or viral particles. A major benefit of the Western blotting technique is that the high affinity antibody–antigen reaction allows antigen detection in the presence of a large excess of other proteins.

Preparation of hydrophilic proteins, such as cytoplasmic or serum components, is usually simple and requires only addition of a detergent to blood or lysis of cells with a non-denaturing detergent and separation of the soluble proteins from cell debris and nuclei. However, it may be necessary to test several different procedures for release of hydrophobic or amphiphilic proteins from membrane-bound compartments or viral particles. A basic procedure for extraction of cytoplasmic and membrane proteins by cell lysis with detergents is described in *Protocol 4*. Different lysis buffers can be rapidly tested for their efficiency in antigen solubilization and denaturation using the dot immunoblotting technique.

Protocol 4. Extraction of cytoplasmic and membrane proteins by cell lysis with detergents

Solutions
- Lysis buffer with Triton X-100: 0.15 M NaCl, 0.02 M sodium phosphate buffer pH 7.60, 2 mM EDTA, 1% v/v Triton X-100 (the Boehringer–Mannheim brand is suitable) in distilled deionized water.
- Lysis buffer with Nonidet P-40: 0.15 M NaCl, 0.01 M Tris–HCl pH 7.40 (at 25 °C), 2 mM EDTA, 1% v/v NP-40 in distilled deionized water.
- Lysis buffer with CHAPSO: 0.15 M NaCl, 0.02 M sodium phosphate buffer pH 7.60, 5 mM EDTA, 10 mM CHAPSO in distilled deionized water.

The lysis buffers should be sterilized by autoclaving and kept refrigerated in dark bottles for less than 1 month. Immediately before use add a protein carrier such as 1 mg/ml BSA and protease inhibitors (20 µg/ml ovatrypsin inhibitor, 50 µg/ml aprotinin, 3 µg/ml PMSF and 3 mg/ml benzidine), and 1 mg/ml N-ethylmaleimide to prevent oxidation of –SH groups in protein molecules.

In contrast to Tris–HCl buffers, the pH of phosphate buffers varies much less with temperature and therefore phosphate-buffered solutions are more suitable for use in procedures such as preparation, storage and electrophoresis of protein extracts.
- PMSF stock solution: 3 mg/ml in 100% isopropanol, stable for at least 9 months at room temperature.
- BSA stock solution: 100 mg/ml in sterile PBS buffer. Store in 1 ml aliquots at −20 °C. Can be refrozen.
- Merthiolate stock solution: 20% in sterile PBS buffer. Store in a dark bottle in refrigerator.

Cell lysis procedure

1a. *Cells from fresh tissues.* Cut the tissue into small pieces and tease it apart using two needles, or crush the tissue between two frosted glass slides. Gently release the cells into a cold protein-free medium (PBS, DBSS or RPMI). Transfer the cells into a plastic tube on ice (capped Corning tubes are convenient) using a pipette and let them sit for 5–10 min. Separate the single cells from sedimented cell clumps and debris by pipetting into a new tube. This procedure is suitable for soft tissues including brain, liver, kidneys, pancreas, spleen, or bone marrow. Other tissues (e.g. muscle) may require different techniques for the release of resident cells. Next step: 2.

Protocol 4. *Continued*

1b. *Cells growing in suspension.* Cells from body fluids (blood, lymph, spinal fluid, etc.) or long-term suspension cultures (e.g. lymphocytes or transformed fibroblasts) can be used directly for protein extraction in this procedure. Next step: 2.

1c. *Attached cells in culture*, such as fibroblasts, astrocytes or epithelial cells. Remove dead cells and debris by gentle washing with warm protein-free medium. Estimate the cell density. Do not detach the cells by trypsin–EDTA since such treatment would alter the cell membrane composition. Instead, wash the cell layer directly in the culture vessel by three rinses (25 ml each) of warm protein-free medium. Carefully remove all the medium after the last wash. Then place the culture dish on ice, overlay cells with a lysis buffer and gently rock on ice for 15–30 min to complete the lysis. Usually at least 10 ml of a lysis buffer is necessary to lyse 10^8 cells in monolayer. Next step: 4.

2. Assess the cell concentration and viability and remove dead cells if necessary (the presence of <5% dead cells is acceptable). Contaminating red blood cells can be lysed by a hypotonic shock. Wash the cell suspension three times in cold protein-free medium using about 100 ml per 10^8 cells each time. After the last centrifugation, carefully remove all the medium above the cell pellet and keep the cells on ice.

3. Vortex the cell pellet very briefly (1–3 sec), return the tube to ice, add lysis buffer and mix by gentle pipetting. Do not vortex in order to avoid the disruption of cell nuclei and the release of nucleic acids. Use 1 ml of lysis buffer per 0.5 to 1×10^8 cells. Allow the lysis to proceed for 15–30 min on ice, on a rocker.

4. Place the cell lysate into a sterile Eppendorf tube(s) and separate extracted proteins from the insoluble material (cell nuclei, cytoplasmic and membrane debris) by centrifugation in a microcentrifuge at 4 °C, for 30 min. Label a new sterile Eppendorf tube with a non-smear marker (e.g. VWR brand) and chill it on ice. Transfer the protein extract (the supernatant) into the prechilled tube and place it immediately at -70 °C, unless the preparation will be used within 1 h. The pellet is usually very small and white. The protein extract has the viscosity of an aqueous solution: a higher viscosity indicates undesirable contamination with DNA.

5. When using a frozen protein extract, transfer the stock preparation from -70 °C to a microcentrifuge shortly before use. Spin at 4 °C for 15 min—the sample will thaw during this time and degraded material will pellet. After removing an aliquot, refreeze the stock extract immediately in the same tube. Most protein extracts remain stable for several months when stored at -70 °C if repeated thawing can be avoided. The molecular integrity of individual proteins in a frozen preparation should be established experimentally, however.

Selection of the antigen-solubilizing agent should be based on antigen binding properties of the available specific antibody (*Table 1*). If the antibody can recognize a denatured form of the antigen, then strong ionic detergents or chaotropic agents (e.g. SDS or urea) can be included in the extraction procedure when non-denaturing tensids alone fail to solubilize the antigen. However, SDS and similar reagents should be avoided during lysis of whole cells since SDS disrupts the cell nucleus and the released DNA makes the lysate very viscous and difficult to analyse.

If the only available probe is a monoclonal antibody specific for a structural epitope on the native antigen molecule, then protein denaturation should be minimized during sample preparation. These antigens should be extracted only with non-denaturing detergents specifically purified for protein chemistry. Pure synthetic tensids with defined, reproducible properties were found to be superior to industrial detergents (e.g. Triton X-100 or Nonidet P-40) and natural products (e.g. cholate salts) in effective solubilization of lipid bilayers without altering the structure of released proteins (26). The compounds octylglucoside, CHAPS, CHAPSO and Zwittergents 310, 312 are among the most suitable non-denaturing detergents available commercially. The native protein extract should be kept on ice, never vortexed or vigorously mixed, and should be analysed immediately to avoid denaturation upon storage. Often, mAbs specific for native antigens can only be used for antigen detection by the dot immunobinding technique (*Table 1*).

In general, protein samples can be successfully stored frozen at −70 °C in small aliquots to avoid repeated thawing, or refrigerated in the presence of an anti-microbial agent such as 0.02% merthiolate. Antigen preparations are usually stable for several months or even years when stored properly. However, routine testing of epitope stability and comparison between stored and fresh samples is recommended.

6. Preparation of antibodies

Both polyclonal and monoclonal antibodies have been successfully used for characterization of specific antigens in Western blotting. The use of either type of antibody is, however, associated with certain advantages and limitations.

A monoclonal antibody will bind specifically to only one epitope, with a particular affinity. In general, the derivation of hybridomas for use in protein blotting involves both considerable skill and no small measure of serendipity. Some monoclonal antibodies bind to stable epitopes on denatured antigens with high affinity and therefore yield excellent probes for Western blots. Such mAbs allow very sensitive and highly specific detection of blotted antigens (*Figure 4*). A large number of hybridoma cell lines producing well characterized mAbs to different antigens can be obtained from the American Tissue Culture Collection. However, many mAbs raised against whole cells, native proteins or intact viral particles fail to bind to the denatured antigens on Western blots. Using dot

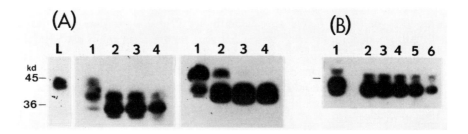

Figure 4. Study of epitope stability with a monoclonal antibody to a class I MHC antigen during SDS denaturation, reduction, deglycoslyation and removal of sialic acid residues. Membrane and cytoplasmic proteins were prepared from BCL_1 B lymphoma cells by extraction with 1% Triton X-100 (10^8 cells/ml). **(A)** The effects of SDS, 2-mercaptoethanol and removal of N-linked glycans on recognition of the D^d antigen by the 7.2.14 mAb. 0.1% SDS (non-reduced samples, **left panel**) or 0.1% SDS plus 1% 2-mercaptoethanol (reduced samples, **right panel**) were added to the protein extract and the samples were boiled for 1 min. Aliquots of the non-reduced and reduced samples were digested with Endoglycosidase F at 37 °C for 12 min (**lane 1**), 120 min (**lane 2**), 5 h (**lane 3**) or 25 h (**lane L** and **4**). No EndoF was added to the sample in Lane L. **(B)** The effect of removal of sialic acid residues on recognition of the D^d antigen by the 7.2.14 mAb. The protein extract was treated with neuraminidase Type X at 37 °C for 1 min (**lane 2**), 6 min (**lane 3**), 12 min (**lane 4**), 24 min (**lane 5**) or 60 min (**lane 6**). No neuraminidase was added to the sample in **lane 1**. After the treatments, the samples (5×10^6 cell equivalents each) were resolved in non-reducing 10% SDS-PAGE gels and electroblotted on nitrocellulose filters. The H-$2D^d$ antigen was detected using the rat anti-D^d mAb 7.2.14 (31) followed by mouse anti-rat kappa mAb RG7/9.1 (34), and [^{125}I]protein A (2×10^5 c.p.m./ml in Blotto-T). Neat hybridoma supernatants were used to prepare the 7.2.14 and RG7/9.1 probing Blotto-T solutions. Blots were washed overnight with Blotto-T, T and then rinsed with the LiCl/NP-40 solution before autoradiography (48 h exposure).

immunobinding, the potential utility of a mAb for antigen detection in Western blotting can be rapidly assessed by testing its reactivity against the antigen in different denaturation states (*Table 1*).

Polyclonal antibodies often provide greater sensitivity for antigen detection by virtue of multiple epitope recognition involving high affinity antibody molecules in antisera. Sometimes, however, probing of Western blots with an antiserum results in visualization of multiple bands which are difficult to interpret, especially if no preliminary information about the size of a specific antigen is available. Non-specific bands are not usually detected when monoclonal antibodies are employed as probes.

Unlike a sample of a monoclonal antibody, an antiserum preparation is a reagent of a limited quantity and unique quality. Consequently, each antiserum preparation derived from different bleedings of the same animal or independent animals immunized with the same antigen has to be individually titrated to ensure reproducible results in antigen detection. On the other hand, the unique immunochemical properties of a monoclonal antibody are maintained upon cloning of the antibody secreting cells. Therefore, most monoclonal antibodies

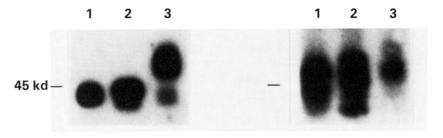

Figure 5. Comparison between a monoclonal and a polyclonal antibody in detection of class I MHC antigens on Western blots. Affinity-purified H-2K^d (**lane 1**), K^b (**lane 2**) and K^k (**lane 3**) antigens were resolved in a reducing 8% SDS-PAGE gel and electroblotted on a nitrocellulose filter. The H-2K antigens were detected by probing with either the rat anti-H-2D^d mAb 7.2.14 (31) (5 μg/ml in Blotto-T) followed by mouse anti-rat kappa mAb RG7/9.1 (34) (5 μg/ml in Blotto-T) and [^{125}I]protein A (2 × 10^5 c.p.m./ml in Blotto-T), in the **left panel**, or with a polyclonal rabbit antibody to purified H-2K antigens (1:300 diluted antiserum in Blotto-T) followed by the same solution of [^{125}I]protein A (**right panel**). The blots were washed overnight in Blotto-T, T before autoradiography (26 h exposure).

can be prepared in virtually unlimited amounts as hybridoma culture supernatants or in the form of ascites fluid.

6.1 Monoclonal antibodies

Supernatants of antibody-secreting hybridoma cells can be often used directly in probing solutions for Western blot detection, without any further purification of the mAb (*Section 9.2*). Serum proteins present in the cell culture media do not appear to interfere with the antibody–antigen reaction on the blot. Sufficient mAb concentrations (1–5 μg/ml) can usually be obtained in supernatants of nearly confluent hybridoma cultures (seeded at about 10^6 cells/ml) collected after 3–5 days of growth. After filter sterilization and addition of 0.02% merthiolate, the supernatants can be stored refrigerated for several months. A working concentration of the mAb should be established for individual supernatant batches. Dot immunobinding provides a rapid way of titrating many antibody samples: Small nitrocellulose strips with dots representing three serial dilutions of the antigen are separately incubated in 1 ml of Blotto-T (16) prepared from the tested mAb preparations, or a pretitrated sample of the mAb. Afterwards, all the strips are pooled for incubation with [^{125}I]protein A.

Monoclonal antibodies can also be prepared using serum-free media, such as Nutridoma, for the hybridoma growth. Usually, a large semi-confluent culture of hybridoma cells (e.g. 10^8 cells in 100 ml of media) is switched into the Nutridoma medium and grown to saturation for 3–6 days. The supernatant is then collected, concentrated 50–100 times using the Amicon concentrator, dialysed against PBS and refrigerated after addition of 0.02% merthiolate. Alternatively, a large amount of a specific mAb can be obtained in ascites fluid (28). Freshly collected ascites fluid, usually containing 2–20 mg mAb/ml should be cleared by a short high-speed centrifugation, titrated and stored in small aliquots at $-70\,°C$.

Various techniques for purification of monoclonal antibodies were compiled recently (28). Commonly used procedures include:
- protein A–Sepharose affinity chromatography suitable for all rabbit, hamster and mouse and most human IgG subclasses, and for γ_1, γ_{2b} and γ_{2c} isotypes of rat IgG (29).
- ion-exchange chromatography (rat IgG_{2a} mAbs).
- euglobuline precipitation followed by Sephadex G200 gel filtration (mAbs of the IgM class) (30).

The recently introduced protein G–Sepharose chromatography, using a recombinant protein G from which the albumin-binding domain was removed, allows efficient purification of IgG antibodies from many species, including all of the γ subclasses from human, mouse and rat, which do not bind protein A. Starting materials suitable for mAb purification by these methods include ascites fluid preparations and concentrated Nutridoma supernatants. The total protein content (by absorbance at 280 nm), the specific immunoglobulin concentration (by ELISA) and the biochemical activity (by dot immunobinding) should be determined in a preparation of a purified mAb. Purified mAbs should be stored in PBS at the antibody concentration of at least 1 mg/ml, frozen in small aliquots at $-70\,°C$ or refrigerated in the presence of 0.02% merthiolate. Sodium azide should not be added to antibody preparations intended for enzymatic radioiodination or procedures involving the use of conjugated enzymes.

When only a limited amount of a specific mAb is available it should be used in the one-step probing approach instead of the standard multi-step sandwich probing. For example, in our experience, approximately 1 mg of a primary mAb is required to probe fifty 150 cm^2 filters by the standard multi-step procedure whereas only about 50 μg of the mAb is needed for the same work when using ^{125}I-labelled primary mAb in the one-step treatment.

6.2 Polyclonal antibodies

Polyclonal antibodies prepared in rabbits, guinea pigs, rats, and mice have been commonly used to probe Western blots in conjunction with radioiodinated protein A as a signal-generating reagent. Most IgG subclasses in sera of these species bind protein A with a high affinity. If the available antibody does not bind protein A well, for example goat or avian antisera, or amplification of antigen detection is required, then a secondary antibody can be used instead of protein A.

The overall immunoglobulin concentration and the titre of specific immunoglobulins should be established in all unique antisera. In good antisera, specific antibodies constitute a large portion of the total serum immunoglobulin. Antisera should always be pretested and used at the highest dilution which still yields a positive signal, in order to minimize non-specific binding.

7. Labelling of purified antibodies and protein A with ^{125}I

Autoradiography of ^{125}I-tagged proteins appears to be still the most sensitive detection technique available in Western blotting.

The radioiodinated reporter molecule is usually a common secondary reagent such as protein A, protein G, or an anti-immunoglobulin antibody. It should be remembered that the use of a polyclonal secondary antibody prior to incubation with the labelled reporter protein will greatly amplify the antigen signal but could also significantly increase background signals. In general, when the secondary reagent is a monolonal rather than a polyclonal antibody, the degree of the final signal amplification is lower but the signal is highly specific. In our experiments (31–33) involving several different primary rat mAbs, good results were obtained using radioiodinated mouse anti-rat kappa chain mAb RG 7/9.1 (34) as the reporter molecule (*Figure 6*).

Excellent results can also be obtained with a radioiodinated primary antibody as the signal-generating reagent (*Figure 7*). The use of a labelled primary antibody provides the advantage of rapid blot processing, high sensitivity and requires about 20-times less of the antibody than multi-step probing procedures.

To assure specificity and sensitivity of probing, only highly purified protein should be used for radiolabelling. Protein A, protein G and various antibodies are available commercially in a highly purified form. Alternatively, a specific immunoglobulin has to be extensively purified from a crude antibody preparation such as ascites fluid, concentrated hybridoma supernatant or antiserum (*Section 6*).

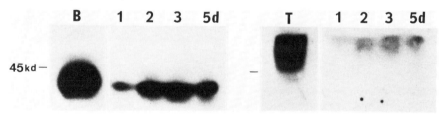

Figure 6. Detection of the H-2D^d antigen or the IL-2 receptor in the same protein samples using different primary antibodies and common secondary reagents. Membrane and cytoplasmic proteins were prepared from BCL$_1$ B lymphoma cells (**lane B**), splenic T cells activated with concanavalin A for 2 days (**lane T**), and small splenic B cells activated with 10 µg/ml *E. coli* LPS and 50 µg/ml dextran sulphate for 1, 2, 3 or 5 days (**lanes 1, 2, 3, 5d**), by extraction with 1% Triton X-100. The protein samples (5 × 10^6 cell equivalents each) were resolved in a non-reducing 10% SDS-PAGE gel and electroblotted on a nitrocellulose filter. The filter was cut into two panels and probed. In the **left panel**, the H-2D^d antigen was detected using the rat anti-D^d mAb 7.2.14 (31), and the p55 IL-2 receptor was detected with the rat mAb 7D4 (38) in the **right panel**. Following probing with the specific primary antibodies (neat hybridoma supernatants in Blotto-T), both panels were incubated together in the mouse anti-rat kappa mAb RG7/9.1 (34) (5 µg/ml in Blotto-T) and [^{125}I]protein A (2 × 10^5 c.p.m./ml in Blotto-T). The blots were washed overnight in Blotto-T, T before autoradiography (3-day exposure).

Detection of macromolecules

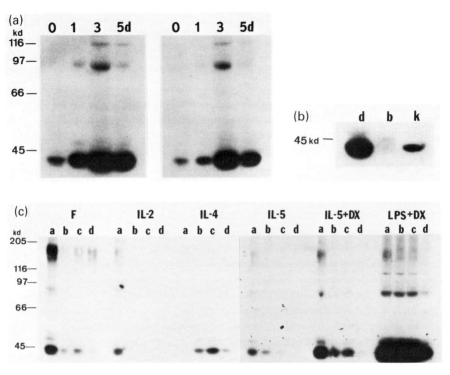

Figure 7. Parallel Western blotting analysis of related groups of protein samples. Splenic B cells were purified by depletion of T cells, red blood cells and adherent cells, and separated into four subsets using a discontinuous Percoll density gradient: enlarged B cells banded at Percoll densities 1.058 and 1.070 g/ml whereas the small resting B cells had densities 1.076 and 1.084 g/ml. The purified B cells were subjected to various treatments described below and then lysed with 1% Triton X-100 to prepare membrane and cytoplasmic proteins (33). Samples of the protein extracts (3×10^6 cell equivalents each) were resolved in non-reducing 8% SDS-PAGE gels and electroblotted on nitrocellulose filters. The H-2D^d antigen was detected by one-step probing with ^{125}I-7.2.14 mAb specific for the D^d protein (31) (2×10^5 c.p.m./ml in Blotto-T). The blots were probed overnight at room temperature, then washed with Blotto-T,T for 8 h and exposed to XAR-5 film with intensifying screen at $-70\,°C$ for 26 h. **(a)** Time course of H-2D^d antigen expression during activation of two subsets of small B lymphocytes. Purified small B cells banding at Percoll densities 1.076 g/ml (**left panel**) and 1.084 g/ml (**right panel**) were stimulated with 10 µg/ml *E. coli* LPS for 0, 1, 3 or 5 days, as indicated. **(b)** Microheterogeneity of the 7.2.14 mAb epitope structure in the molecules of H-2D antigens of the d, b and k mouse MHC haplotypes. Purified small B cells (Percoll density 1.084 g/ml) were prepared from mice of different H-2 halpotypes: BALB/c (**lane d**), BALB.B (**lane b**) and BALB.K (**lane k**), and stimulated with 10 µg/ml *E. coli* LPS for 3 days. **(c)** Expression patterns of the H-2D^d antigen in four different subsets of B lymphocytes during stimulation with lymphokines and mitogens. Purified splenic B cells, banded at Percoll densities 1.058 g/ml (**lane a**), 1.070 g/ml (**lane b**), 1.076 g/ml (**lane c**) and 1.084 g/ml (**lane d**), were either lysed immediately (**panel F**, fresh cells) or cultured for 3 days in the presence of 10% FCS and different stimuli: 200 U/ml rIL-2 (**panel IL-2**), 10 pM IL-4 (**panel IL-4**), 10 pM IL-5 (**panel IL-5**), 10 pM IL-5 plus 50 µg/ml dextran sulphate (**panel IL-5+DX**), or 10 µg/ml *E. coli* LPS plus 50 µg/ml dextran sulphate (**panel LPS+DX**).

In order to maintain the molecular integrity of radiolabelled proteins, gentle modification conditions should be used such as those provided by the Enzymobead-catalysed reaction. The Enzymobeads, available from Bio-Rad Laboratories, are a solid-phase enzymatic radioiodination reagent consisting of immobilized preparations of lactoperoxidase and glucose oxidase. On addition of glucose and [^{125}I]NaI to the Enzymobead suspension, free iodine is generated which reacts with the protein present in the solution. After quenching the reaction, the remaining free iodide is removed from the labelled protein by a small scale gel filtration (*Protocol 5*). This radioiodination method is very efficient yet it does not perturb the protein structure. We have used the Enzymobead reagent routinely for labelling immunoglobulins of various classes, prepared by affinity chromatography or euglobulin precipitation, and protein A. Using 50 μg of a purified immunoglobulin, or protein A, and 1 mCi of [^{125}I]NaI in a typical reaction (*Protocol 5*), we prepared radioiodinated proteins with specific activity of 1×10^7–3×10^7 c.p.m./μg. The radiolabelled protein should be stored refrigerated in a stabilized PBS solution. Radioiodinated compounds should only be handled by trained personnel. In our experiments, ^{125}I-labelled proteins (including protein A, rat IgM mAbs, rat IgG mAbs, mouse IgG mAbs and rabbit polyclonal antibodies) retained their molecular structure and full biochemical activity for over two months when stored as described above, at concentrations of about 10^9 c.p.m./ml. We have also successfully re-used Blotto solutions containing diluted radiolabelled proteins (about 5×10^5 c.p.m./ml) (*Section 9.3*).

Protocol 5. Preparation of radioiodinated proteins using Enzymobeads

Reagents

All reagents must be free of sodium azide and thiocyanate to avoid inhibition of the lactoperoxidase.

- Enzymobeads (immobilized lactoperoxidase glucose oxidase available from Bio-Rad Laboratories, Richmond, CA): rehydrate the reagent with sterile distilled water as recommended by the manufacturer. Store in 25-μl aliquots (sufficient for one reaction) at −20 °C. Do not refreeze.
- Sterile 0.2 M phosphate buffer pH 7.2. Store in small aliquots at −20 °C. Can be refrozen.
- Sterile 1% β-D-glucose in distilled water. If the crystalline β-D-glucose is not available, use D-glucose and allow mutarotation to proceed at room temperature overnight. Store in small aliquots at −20 °C. Can be refrozen.
- Protein samples.
    (a) 10 mg/ml affinity purified protein A (available commercially) in sterile PBS buffer. Store in 50–500 μg aliquots at −20 °C or −70 °C. Can be refrozen.
    (b) 2–10 mg/ml purified immunoglobulin (mAb or a polyclonal antibody) in sterile PBS buffer. Store in 50–100-μg aliquots (for one reaction) at −70 °C. Do not refreeze.

Detection of macromolecules

Protocol 5. *Continued*

(c) Carrier-free stabilized [^{125}I]NaI solution, radioactive concentration about 500 mCi (18.6 GBq)/ml (the IMS.300 reagent from Amersham is suitable). Store at room temperature.

(d) Carrier protein: 100 mg/ml BSA in sterile PBS buffer. Store in small aliquots at $-20\,°C$. Can be refrozen.

- 10% Tween-20 detergent in sterile distilled water. Store at room temperature in a dark bottle.
- Biogel P-10 suspension (a gel filtration matrix for removal of unbound NaI): rehydrate 4 g of dry gel with 48 ml of sterile PBS buffer containing 0.05% Tween-20. Leave the matrix and buffer in a 50-ml Corning tube on a rocker or rotator at room temperature overnight. Store at $+4\,°C$, in the presence of 0.02% merthiolate. Vortex before use.

Preparation of a prespun Biogel P-10 minicolumn

The minicolumn should be prepared about 5 min before use to avoid dehydration of the gel bed.

1. Take one 0.9 ml and one 1.8 ml Eppendorf test tube. Puncture a very small hole in the bottom of the smaller tube with the tip of a 22-guage needle.
2. Pipette 0.9 ml of the Biogel P-10 suspension into the smaller tube. Use the opened larger tube as a carrier for the smaller tube.
3. Settle the gel bed by centrifugation of the stacked small and large tubes in a table-top clinical centrifuge (MSE), 800 r.p.m. ('speed 5'), at room temperature for 3 min. Discard the excluded buffer in the large tube.
4. The minicolumn, a 0.5-ml gel bed in the smaller tube, is ready to use for gel filtration of a 100–200-μl sample.

Radioiodination procedure

1. Mix 25 μl of the rehydrated Enzymobead reagents with 30 μl of the phosphate buffer, 25 μl of the glucose solution and 5–25 μl of the protein sample (50 μg of protein A or 50–100 μg of an immunoglobulin) in a sterile Eppendorf tube at room temperature.
2. In a chemical hood equipped for work with radioiodine, add 1 mCi of the [^{125}I]NaI solution (2–5 μl) to the reaction mixture, mix carefully all the reagents and incubate at room temperature for 25 min.
3. Prepare a prespun minicolumn of Biogel P-10: just before the end of the incubation time. Prechill a small Eppendorf tube with 2 μl of each of the BSA and the Tween-20 solutions.
4. Terminate the radioiodination reaction: pellet the Enzymobeads in a minicentrifuge for 1 min, rapidly remove the supernatant into the tube with BSA plus Tween-20 on ice and mix. Immediately transfer this mixture on to the prespun minicolumn.

Protocol 5. *Continued*

5. Separate the unbound [^{125}I]NaI from the iodinated protein: centrifuge the Biogel P-10 minicolumn in a table-top clinical centrifuge (MSE) at 800 r.p.m. for 1 min. Discard the small tube with the Biogel into radioactive waste. Add 100 µl of PBS buffer to the eluate in the larger tube to obtain a stabilized solution of the radioiodinated protein in the presence of about 1 mg/ml BSA and 0.1% Tween-20.

6. Spot 1 µl of the radioiodinated solution ($\sim 0.5\%$) on to a small piece of filter paper, place it in a carrier tube and count the radioactivity in a γ-counter (for 30 sec). Calculate the radioactive concentration (c.p.m./µl) and specific activity (c.p.m./µg) of the radiolabelled protein. Using this procedure we have prepared radioiodinated proteins with specific activites of about $1-3 \times 10^7$ c.p.m./µg.

7. Store the ^{125}I-labelled proteins in a lead container at 4 °C for up to 2 weeks. Some labelled immunoglobulins may be less stable. Handle all ^{125}I-containing material in double gloves, in an appropriately shielded chemical hood.

8. Gel electrophoresis and electrophoretic transfer

Identification and analysis of homologous antigen bands in different protein samples can be greatly facilitated if electrophoresis and probing of the samples are carried out under identical conditions. It is therefore advantageous to plan Western blotting experiments in such a way that the maximum capacity of the available equipment is utilized so that a large group of samples can be processed together. For example, using two double-sided vertical gel boxes and a double-cassette electrophoretic transfer cell, four slab gels can be run and transferred simultaneously allowing parallel analysis of up to 80 protein samples. This approach is particularly helpful in time-course experiments of antigen accumulation (*Figure 7a*), tissue distribution, molecular polymorphism (*Figure 7b*), or induction by different stimuli (*Figure 7c*). Four or more protein blots can be probed and washed together, in the same solution, in one container, without compromising the specificity of the antigen detection or increasing non-specific background.

If the antigen epitope is stable in the presence of SDS, the antigen sample can be separated using a standard SDS-PAGE gel. In case of antigens with SDS-labile epitopes, a non-denaturing detergent (Nonidet P-40 or CHAPSO) can be substituted for SDS. Alternatively, SDS-denatured antigens can be partially renatured directly in the gel after electrophoresis (*Section 8.2*).

8.1. Preparation of antigen samples for electrophoresis

The selection of optimal sample concentration and conditions for sample preparation should be based on test experiments systematically examining factors which affect the antigen epitope stability (*Table 1*) with regard to electrophoresis and electroblotting, such as detergents (0.1% or 0.5% SDS), 2-mercaptoethanol (2% v/v), freezing–thawing and heat (22 °C and 100 °C). The epitope stability and the actual concentration of antigen in the sample can be rapidly assessed using dot immunobinding. Considering an extract of cellular proteins containing 10^8 cell equivalents per ml (*Protocol 4*), the lysate of up to 7×10^6 cells (70 μl) can be analysed in one lane of a standard 1.5 mm thick SDS-PAGE gel. This sample amount should allow detection of abundant cellular proteins expressed at the density of at least 10^3 molecules per cell. For example, 50 μl of this cell lysate would contain about 5 ng of a 60-kd antigen expressed at the density of 10^4 molecules per cell and, using a radiolabelled probe, this antigen amount would be readily detectable after 16–24 h of autoradiography. In our experiments, 10–50 μl of such lysate (i.e. $1-5 \times 10^6$ cell equivalents) were suitable for sensitive detection of different murine proteins, including different class I MHC and class II MHC antigens, IL-2 receptor, T cell receptor, transferrin receptor and membrane IgM.

Detection of a less abundant antigen may require concentration of the antigen prior to Western blotting, for instance by affinity chromatography or immunoprecipitation. Alternatively, a different method for the initial protein sample preparation can be used which would yield a more concentrated antigen solution, such as Triton X-114 extraction of hydrophobic and amphiphilic cellular proteins (35).

Frozen stock antigen preparations should be transferred from -70 °C only shortly before removing a sample for electrophoresis and immediately refrozen in the same tube (*Protocol 4*). Unless the antigen epitope is heat-labile, further manipulations of the sample can be carried out at room temperature.

Preparation of non-reduced samples involves mixing the desired amount of the antigen extract with 1/10 of the volume of a $10 \times$ concentrated sample buffer (WSB buffer in *Protocol 6*). The WSB buffer should be mixed by tapping the tube gently. Vortexing should be avoided since it denatures proteins and causes foaming of the solution which makes it difficult to load the sample on gel afterwards. It is not necessary to increase temperature of the sample above room temperature for SDS-mediated protein denaturation to occur.

In the case of reduced samples, neat 2-mercaptoethanol solution should be added to the antigen extract together with the WSB buffer so that the final concentration in the sample is 2% v/v. The reduced samples should then be boiled for 5 min to facilitate reduction of the S–S bonds in protein molecules, chilled on ice briefly and spun in a minicentrifuge for 2 min to remove denatured material before loading on to the the gel.

Protocol 6. Solutions used in discontinuous Laemmli SDS-PAGE electrophoresis and electroblotting

Use only distilled deionized water and electrophoresis purity reagents in preparation of all the solutions.

Protocol 6. Continued

- Concentrated acrylamide–bis acrylamide solution: 30% w/v acrylamide and 0.8% w/v bis. Sterilize by filtration, store refrigerated in a dark bottle for up to 1 month.
- Concentrated stacking gel buffer: 2 M Tris–HCl, pH 6.8 (at 25 °C). Sterilize by autoclaving, store refrigerated after opening.
- Concentrated resolving gel buffer: 1.5 M Tris–HCl, pH 8.8 (at 25 °C). Sterilize by autoclaving, store refrigerated after opening.
- 10% SDS. Store at room temperature.
- 10% w/v ammonium persulphate (APS). Make fresh daily or store in 0.1-ml aliquots at -20 °C; do not re-use.
- TEMED. Store refrigerated in a dark bottle for up to 6 months.
- 10% resolving gel (30 ml): mix 7.5 ml of resolving gel buffer, 12 ml of acrylamide–bis solution, 0.3 ml of 10% SDS, 10 ml of water. De-gas the solution. Add 0.2 ml of 10% APS, 20 μl of TEMED and use immediately.
- 4% stacking gel (10 ml): mix 0.625 ml of stacking gel buffer, 1.3 ml of acrylamide–bis solution, 0.1 ml of 10% SDS, 7.9 ml of water. De-gas the solution. Add 50 μl of 10% APS, 10 μl of TEMED and use immediately.
- 4 × Tris–glycine electrophoresis buffer (4 × Laemmli buffer): 0.1 M Tris–base and 0.768 M glycine, pH 8.8 (at 25 °C). Sterilize by autoclaving, store refrigerated for up to 1 month. Immediately before use, dilute to 1 × concentration with water and add 0.1% SDS using a concentrated SDS solution.
- 10 × non-reducing Western sample buffer (10 × WSB): 0.5 M Tris–HCl pH 6.8 (at 25 °C), 5% w/v SDS, 50% v/v glycerol and 0.1% w/v bromophenol blue dye. Sterilize by filtration. Store at -20 °C for at least 1 year, or refrigerated for up to 3 months. Warm up to room temperature before use. Use 1 μl per 10 μl protein sample, mix gently—do not vortex—and incubate at room temperature for 10–30 min before loading on gel.
- 10 × reducing Western sample buffer: mix 2 μl of neat 2-mercaptoethanol and 10 μl 10 × WSB, use 1.2 μl per 10 μl of protein sample. Boil the sample for 5 min, chill on ice and spin in a microcentrifuge for 2 min before loading on gel.
- Tris–urea solution: 0.05 M NaCl, 4 M urea, 10 mM Tris–HCl pH 7.0 (at 25 °C), 2 mM EDTA. Sterilize by autoclaving, store at room temperature. Immediately before use, cool to 4 °C and add 0.1 mM dithiothreitol.
- Towbin buffer for electrophoretic gel transfer (4 litres): mix 2.4 litres of water with 0.8 litres of methanol and 0.8 litres of 4 × Laemmli buffer. Make fresh for each use. Before transfer, cool to 4 °C.

Total protein staining on nitrocellulose using Amido black

- Amido black (AB) destaining solution: 25% v/v isopopanol, 10% v/v acetic acid in water.
- AB staining solution: 0.1% w/v Amido black dye in the destaining solution.
- Molecular weight protein standards: the SDS-6H marker from Sigma (St Louis, MO) covers a 29–205-kd range while the Dalton Mark VII-L marker

Detection of macromolecules

Protocol 6. *Continued*

from Sigma is useful for a 14.2–66-kd range. The protein markers are reduced before electrophoresis and 1 µg per band is usually used. Run the marker in a side lane of a Laemmli SDS-PAGE gel and electroblot on a nitrocellulose filter. Total protein staining is not suitable for nylon membranes.

Staining procedure

1. Cut off the marker lane strip from a nitrocellulose blot immediately after transfer.
2. Soak the strip in the AB destain solution for 2 min.
3. Soak the strip in the AB stain solution until blue protein bands become visible (1–3 min).
4. Soak the stained strip in the AB destain solution until the background becomes clear (agitate for 10–30 min).
5. Rinse the strip in water, store wet in plastic wrap to compare later with protein bands detected on the blot. If allowed to dry, the membrane would shrink.

8.2 Gel electrophoresis

SDS-PAGE gels can be used for electrophoretic separation of a protein sample, if the SDS-denatured antigen can still bind the antibody probe. In the discontinuous Laemmli system, the porosity of a resolving gel can be varied over a wide range to meet separation requirements of the specific antigen: 15% (acrylamide- –bis) gel for antigens of mol. wt. below 43 kd, 10% gel for mol. wt. between 15 kd and 70 kd, and 7% gel for mol.wt. up to 250 kd. If the antigen molecular weight is not known in advance, protein samples should be resolved either using two gels: 15% and 7%, or a 5–15% gradient gel.

SDS-PAGE gels should be prepared shortly before electrophoresis, using concentrated stock solutions of acrylamide with bis, gel buffers, SDS and TEMED (*Protocol 6*). Before loading protein samples, the gel should be prerun (e.g. 20 mA for 30 min) to remove unpolymerized acrylamide and charged impurities, which may cause undesirable protein degradation, from the anodic gel edge. Samples should then be loaded using a Hamilton syringe, rinsing the needle between individual samples in the lower buffer reservoir, or using disposable flat slim tips. Preferably, lanes at the gel extremities should be used only for molecular weight markers. It is advantageous to use prestained mol. wt. markers (available from BRL-Gibco in the ranges of 3–43 kd and 14.3–205 kd) which allow immediate assessment of protein separation during electrophoresis and the efficiency of transfer. Alternatively, protein markers can be visualized by total protein staining after blotting on to a nitrocellulose (not nylon) membrane using Amido black (*Protocol 6*). In a gel, reduced samples of proteins and mol. wt. markers must be separated from non-reduced protein samples by an empty lane since horizontal diffusion of 2-mercaptoethanol can cause reduction of proteins in the adjacent lanes during electrophoresis. Empty lanes should be filled by 20 µl of a diluted sample buffer (1xWSB, *Protocol 6*) to prevent distortion of protein migration in adjacent lanes. It is helpful to mark the lanes with a VWR marker on the large glass plate so that this information can be marked over on the blotting membrane following the transfer (use a pencil or a ball-point pen).

Protein samples should be electrophoresed under constant current, at first using a low current (~ 10 mA per gel) until the bromophenol blue dye has migrated about 1 cm into the stacking gel. Then a higher current level can be set (~ 40 mA per gel) to separate proteins rapidly, or a low current level (~ 8–15 mA) can be maintained for about 12–16 h. In the low current electrophoresis procedure, the gel temperature does not increase above the ambient level and bands of proteins, including those of high molecular weight, are well separated, straight and sharp (*Figure 7*). In case of heat-labile antigens, electrophoresis should be carried out at low current in a cold room, or in a gel box with a cooling coil attached to a refrigerated circulating water bath. The choices of actual current levels and the duration of electrophoresis should be optimized empirically. The parameters suggested above refer to the system used in our laboratory: 15 cm × 15 cm × 1.5 mm SDS-PAGE gels, the V16-2 gel box from BRL and a Tris–glycine elctrophoresis buffer (*Protocol 6*).

Immediately after electrophoresis, the gel should be removed from glass plates and transferred. However, in case of antigens with SDS-sensitive epitopes, it is possible to renature the antigens partially, prior to transfer. After electrophoresis, the SDS-PAGE gel is first equilibrated (gently rocked in a dish) in a buffered chaotropic solution (Tris–urea solution; *Protocol 6*) at room temperature for 2–12 h, and then re-equilibrated in a cold 1 × Laemmli buffer (*Protocol 6*) with 0.1 mM dithiothreitol for at least 30 min before transfer. The gel is then electroblotted in cold 1 × Laemmli buffer. Successful recovery of immunoreactivity in SDS-denatured antigens has been reported by several laboratories using this approach (36, 37).

8.3 Electrophoretic transfer

Sensitive antigen detection on Western blots requires highly efficient transfer of the antigen from a gel to an immobilizing matrix. Several types of blotting apparatus are available commercially which allow nearly quantitative transfers of macromolecules by virtue of a high electric field strength generated in the blotting tank. Important accessories provided with an efficient blotting apparatus include grid-type electrodes mounted on removable cards, hinged gel holders and a blotting tank with a cooling core. The grid electrodes generate nearly homogeneous, high current density fields that facilitate a more uniform transfer of proteins. The removable mounted electrodes can be moved closer together for very high field strength applications such as transfer of high molecular weight proteins. The hinged gel holders with clamps, colour-coded sides and foam pads provide a tight contact between the gel and the matrix membrane, and ensure correct orientation of the gel matrix sandwich towards the electrodes in the blotting tank. Relatively large joule heat is dissipated during electroblotting. The amount of thermal energy released depends on the strength of the applied field and the conductivity of the transfer media. Uncontrolled heating during transfer may cause irreversible protein denaturation, deformation of the matrix membrane as well as damage to the apparatus. Consequently, efficient cooling is essential for the electroblotting process. Most blotting tanks are equipped with a cooling coil which can be connected with a refrigerated circulating water bath to maintain an appropriate temperature in the tank. In our laboratory, an alternative procedure was used successfully. We used precooled transfer buffer and carried out the transfer in a cold room, with the tank standing on a magnetic stirring plate to ensure gentle buffer circulation during blotting.

The most commonly used medium for electrophoretic transfer of proteins from SDS-PAGE gels is a Tris–glycine buffer with methanol, originally introduced by Towbin et al. (1) (*Protocol 6*). The Towbin buffer contains 20% v/v methanol which causes dissociation of SDS from SDS–protein complexes, and decreases the gel pore size. Methanol thus promotes renaturation of SDS-denatured proteins while it impedes elution of high molecular weight proteins from gel. The Towbin system can therefore be disadvantageous in case of proteins with alkaline pI which upon removal of SDS become positively charged or neutral and migrate poorly in the buffer. Addition of 0.1% SDS to the Towbin buffer helps to maintain the negative charge of such proteins during transfer as well as to improve the efficiency of elution of large proteins from gel. On the other hand, the presence of SDS in a transfer buffer can compromise epitope stability in some antigens, and it also decreases the efficiency of protein binding to nitrocellulose (but not to nylon) membranes. It is therefore recommended that in special circumstances, such as transfer from non-denaturing gels, blotting of antigens of unusual molecular properties, or the use of particular types of immobilizing matrices, alternative transfer media should be used.

Several different types of immobilizing matrices are available commercially for use in protein blotting. Commonly used matrices include pure nitrocellulose, nylon 66 (Gene Screen from DuPont NEN Products, Nytran from Schleicher & Schuell or Magna Nylon 66 from Micron Separations, Inc.) and cationic nylon (Zeta Probe from Bio-Rad or Zeta-Bind from AFM Cuno). Cationic nylon, a derivative of Nylon 66, is better suited for Western blotting applications than Nylon 66 since the net charge of cationic nylon is not affected by pH of the buffer. Nitrocellulose membranes should be used for initial antigen studies since nitrocellulose has a high affinity and binding capacity for proteins, and unoccupied binding sites can be easily blocked during 30 min pre-incubation in Blotto. In our experience, the 0.2-μm pore size type of nitrocellulose has better protein binding properties than the 0.45-μm type membrane, in particular when dealing with low molecular weight proteins. However, nitrocellulose membranes are less suitable for use with very small or large proteins (<15 kd and >150 kd). In addition, nitrocellulose is relatively fragile and therefore sequential probing with several antibodies can be more difficult. On the other hand, nylon membranes are very durable and bind proteins of a wide range of molecular weights very efficiently even in the presence of SDS or methanol. Because of the high protein binding capacity of nylon, however, blocking of unoccupied sites typically requires an overnight pre-incubation in Blotto.

In order to prevent diffusion of protein bands, transfer should be initiated immediately after electrophoresis. One sheet of the matrix (a nitrocellulose or nylon membrane) cut to the size of the gel, two foam pads (provided for each gel holder with the blotting apparatus) and two sheets of the Whatman 3MM paper cut to the size of the foam pads should be briefly presoaked in the transfer buffer and air should be carefully squeezed out of the wet pads. After electrophoresis, place the gel sandwich on a paper towel (the large glass plate with marked lanes face down), separate the glass plates using a plastic spatula and remove the small glass plate. The fringes (sample wells) on the top of the stacking gel can be trimmed short using a plastic blade (e.g. a ruler). The gel slab should then be carefully overlaid with the wet sheet of matrix. It is helpful to mark the positions of individual samples (i.e. the lane numbers) on the matrix sheet with a ball-point pen. Place one sheet of the wet 3MM paper over the matrix, cover it with a clean glass plate (to facilitate the manipulation of the sandwich) and turn it upside down. Then remove the large glass plate, and cover the gel sequentially with the other wet 3MM paper and a wet foam pad. Place the other pad on the other side of the sandwich. Air bubbles should be carefully removed when the sheets of the matrix and the 3MM paper are laid down in the gel. Clamp the final sandwich (pad–paper–matrix–gel–paper–pad) firmly into the gel holder so that a tight contact between all layers is maintained. Then immediately insert the holder into a blotting tank filled with cold transfer buffer such that the gel faces the anode electrode. This orientation is appropriate for transfer of negatively charged native antigens and SDS–protein complexes. In the case of native basic antigens, the gel holder orientation in the tank should be reversed, and for antigens with unknown pI values, a matrix sheet should be placed on each side of the gel. During the electrophoretic transfer, slow stirring of the buffer by a large magnetic bar helps to remove trapped air bubbles and prevents undesirable thermal and ionic gradients in the blotting tank. If possible, a high current, variable voltage power supply should be used for protein blotting so that a high intensity electric field can be maintained during the transfer. In our experiments using a Hoefer blotting apparatus with a built-in power supply, placed on a magnetic stirring plate in the cold room, transfers were started at the highest available voltage (marked 100% on that instrument) and approximately 0.6 A, and terminated in about 3 h when the current increased to 1.5–2 A. The buffer temperature usually increased from 4 °C to about 30 °C during blotting. After the transfer, the matrix sheet should be removed and immediately placed in Blotto to saturate non-specific protein binding sites.

9. Immunochemical probing of Western blots and autoradiography

In Western blotting, antigens immobilized (blotted) on to a nitrocellulose or nylon membrane are detected through an immunochemical reaction with a specific primary antibody. The antigen–antibody complex on a blot can be visualized either directly, if the primary antibody serves as the signal-generating reporter molecule, or indirectly using a sandwich technique (*Figure 8*). The sandwich techniques involve sequential incubations with several reagents each possessing high mutual affinity and a final reporter molecule. Different reagents have been used in the sandwich techniques including anti-immunoglobulin antibodies, protein A, protein G, and conjugated enzymes (28). ^{125}I-labelled proteins detectable by autoradiography, and enzymes with chromogenic substrates (e.g. horseradish peroxidase, alkaline phosphatases, or glucose oxidase) allowing colorimetric detection, have been commonly employed as reporter molecules in immunoblotting. Biotin-labelled proteins detectable through avidin (streptavidin) coupled to various labels, and antibodies in complex with colloidal gold particles provide alternative means for antigen detection on immunoblots.

The highest level of antigen detection sensitivity can be achieved using a reporter protein molecule labelled with radioiodine. The radionuclide ^{125}I is employed by most, however ^{131}I may also be used if a particularly high sensitivity is desired. Using a ^{125}I-labelled antibody or protein A with specific activity of at

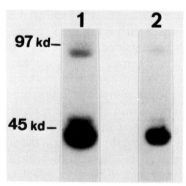

Figure 8. Detection of the H-2D^d antigen using one-step probing with a radiolabelled primary antibody or a sandwich probing procedure with radiolabelled protein A. Membrane and cytoplasmic proteins were prepared from BCL$_1$ B lymphoma cells by extraction with 1% Triton X-100. Two samples of the protein extract (1 × 10^6 cell equivalents each) were resolved in a non-reducing 8% SDS-PAGE gel and electroblotted on a nitrocellulose filter. In **lane 1**, the H-2D^d antigen was detected by probing with ^{125}I-labelled 7.2.14 mAb (1 × 10^5 c.p.m./ml Blotto-T), a specific rat IgM against the D^d protein (31). In **lane 2**, the D^d antigen was visualized by sequential probing with the rat mAb 7.2.14, the mouse anti-rat kappa mAb RG7/9.1 (34) (each mAb at 5 μg/ml in Blotto-T) and ^{125}I-labelled protein A (2 × 10^5 c.p.m./ml in Blotto-T). Both blot strips were washed in Blotto-T, T for 8 h before autoradiography (48-h exposure).

least 3×10^7 c.p.m./μg as a probe, 10–100 pg of antigen per band can be detected by a 48 h autoradiographic exposure at $-70\,°C$ with fast X-ray film and an intensifying screen. In general, the sensitivity of detection techniques with a colorimetric readout is about ten times lower but this can be improved by a subsequent signal amplification, such as silver staining of the immunogold complexes, provided that the introduction of additional reagents does not cause increased background. A large signal amplification (up to 100 times) can be obtained if a polyclonal secondary antibody is used in antigen detection, either as a reporter or as a part of a sandwich complex. The usefulness of this approach is limited by a concomitant amplification of non-specific signals because most anti-immunoglobulin antisera which are commonly used as secondary antibodies react with multiple bands on blots of cellular lysates.

The final detection sensitivity which can be achieved in a particular system of a given antigen and antibody is also dependent on the level of stringency with which the blot can be washed after incubation with specific detection reagents. Stringent washing (in the presence of 0.1% Triton X-100, or 1% NP-40 with 0.5 M LiCl, or 0.1% SDS) can practically eliminate background signals caused by cross-reactive and non-specific protein binding and allow amplification of the specific signal by extended exposure time during autoradiography. However, a stringent wash cannot be employed in blotting systems which involve low affiinity antibodies. Optimally, the level of stringency appropriate for each system of specific reagents should be established experimentally.

9.1 Treatment of blots before probing

An important factor in the control of sensitivity and signal-to-noise ratios in a blotting system is saturation (blocking) of low affinity protein interactions on a blot before incubations with specific protein reagents.

Several different blocking solutions which have been employed in Western blotting contain an excess of a common protein such as bovine serum albumin, ovalbumin, casein or gelatin as the blocking agent. A very simple, inexpensive and versatile blocking solution (Blotto) based on dry non-fat milk (available at grocery food stores) has been introduced by Johnson et al. (27). Blotto can be used for blocking, as a diluent for all the commonly used detection reagents and also for washing of blots. In our experiments, the original Blotto solution composition has been modified by addition of a non-ionic detergent, Tween-20, to reduce non-specific hydrophobic interactions (Blotto-T; *Protocol 7*) and an additional detergent, Triton X-100, was added to the Blotto for the final washing of blots (Blotto-T, T; *Protocol 7*).

Protocol 7. Solutions used in probing of Western blots

Use only deionized distilled water in preparation of all the solutions.
- Blotto-T: 5% w/v non-fat dry milk (available in grocery stores), 0.1%

Protocol 7. *Continued*

Tween-20, 0.01% Anti-foam A (Sigma) in PBS buffer, pH 7.2. Make fresh for each use. It is convenient to use 20 × PBS and 20% Tween-20 solutions in the preparation of Blotto.
- Blotto-T, T: 0.1% Triton X-100 in Blotto-T.
- LiCl/NP-40 solution for stringent washing of blots: 0.5 M LiCl, 0.1 M H_2O Tris–HCl pH 8.0 (at 25 °C), 1% Nonidet P-40 in water. Store at room temperature.
- 20 × PBS buffer: dissolve 47.75 g of anhydrous Na_2HPO_4 in 700 ml of water, add 10.25 g of $NaH_2PO_4 \cdot H_2O$ and allow to dissolve. Then add 175.32 g of NaCl. Adjust the pH to 7.2 with a concentrated NaOH solution. Adjust the volume to 1 litre with water. Sterilize by autoclaving. Store at room temperature.

Blots should be placed into blocking solution immediately after the electrophoretic transfer has been terminated. Rinsing or drying the blot after the transfer is not necessary.

In the case of nitrocellulose membranes, a short incubation in Blotto (30 min) is sufficient for blocking, although the blot can be left in Blotto overnight, if this is more convenient. Nylon membranes, however, require overnight blocking in Blotto due to their high protein binding capacity and net charge. Blocking should be carried out at the same temperature as the subsequent steps for probing and washing (room temperature or 37 °C). During blocking, the membrane should be well covered by solution and agitated. A plastic box with about 200 ml of Blotto, placed on a rocker or an orbital shaker is suitable. During overnight incubations, nitrocellulose membranes should be agitated gently on a rocker to prevent shearing and disintegration of the blot. Efficient blocking and washing requires several changes of generous amounts of the solution. In our procedure designed for simultaneous treatment of four blots, a total of 4 litres of Blotto are required for blocking, dilutions of antibodies and washing the blots.

9.2 Probing of blots with antibodies and protein A

In immunoblotting, antigen detection involves incubation of a preblocked blot with one or more reagents (probes) including a primary antibody specific for the antigen and a reporter molecule allowing final visualization of the antigen. A one-step probing procedure requires only one reagent: the primary antibody labelled with ^{125}I. Multistep (sandwich) probing techniques involve sequential incubations of a blot with several reagents, most commonly with the primary antibody plus a secondary, anti-immunoglobulin antibody then [^{125}I]protein A (*Figure 8*). In between treatments with the individual specific reagents, the blot is washed by vigorous agitation in the blocking solution (Blotto-T, *Protocol 7*), three times for 5–10 min at room temperature.

It is important to avoid the use of excessively high antibody concentrations, in particular with polyclonal antisera, since cross-reactions and low affinity binding to a blot may occur and cause high overall background, non-specific protein detection and other artefacts. Most high avidity mAbs are used at a concentration 5 µg/ml or lower, while high titre polyclonal antisera can usually be diluted at least 1:400 when used as probes. A blocking solution (e.g. Blotto-T, *Protocol 7*) can be used as a diluent for concentrated antibody preparations. We have also found that crude hybridoma supernatants, containing 10% calf serum and the RPMI medium, can be directly used for preparation of antibody probes: the Blotto-T solution (*Protocol 7*) is assembled using a filtered neat culture supernatant in place of PBS plus antibody.

Blots should be incubated in the antibody solution either at 37 °C for 3 h, or overnight at room temperature, in the case of most mAbs and antisera. The use of low avidity antibodies may require different conditions including increased antibody concentration and shorter incubation time. Saturation of a blot with a [^{125}I]protein A solution requires incubation for 2–3 h at 37 °C. Blots should be agitated (rocked) during the incubations to ensure even distribution of the probing solution. In most studies, economical usage of specific antibodies is desirable and therefore a procedure requiring only small volumes of probing solutions is preferred. In our experience, a 200 cm^2 blot can be successfully probed with about 10 ml of solution (Blotto-T plus an antibody) in a heat-sealed bag slightly larger than the blot (Seal-a-Meal bags are available from general drugstores or some biotechnology companies including ASP and BRL). Air bubbles have to be carefully removed before sealing the bag to prevent blank, untreated spots on the blot. After incubation with a blot, the antibody solution can be saved for further use.

In our laboratory, we have prepared ^{125}I-labelled immunoglobulins and protein A using the Enzymobead reagent (*Section 7, Protocol 5*). ^{125}I-labelled protein A is also available commercially, however, at a price vastly exceeding the cost of all the reagents required for its preparation in the laboratory. We have used approximately 10^7 c.p.m. of a radioiodinated antibody or protein A, diluted in 50 ml of Blotto-T, for the probing of one 200 cm^2 blot. Alternatively, we have probed four such blots simultaneously in 100 ml of Blotto-T containing 4×10^7 c.p.m. of the labelled protein. We have re-used the diluted probing solutions, which were refrigerated and contained 0.02% merthiolate, for two or three additional incubations with blots within two weeks.

When preparing a probing solution, add the required amount of the concentrated radioiodinated protein preparation to the bulk of Blotto-T solution in a plastic container, and mix by a gentle agitation before putting the blot(s) into the mixture. Then cover the container with a tightly fitting lid and place it in a lead-lined box to afford shielded handling. Such a box can be easily assembled from a plastic shoe-box lined with trimmed pieces of thin lead shields, available from companies dealing with radiosafety products. Any containers with ^{125}I-labelled material should only be opened and manipulated in a chemical hood with appropriate radioactive shielding. All radioiodinated solutions and dry waste, including the first washing solution after a ^{125}I-labelled probe, should be discarded according to the radiosafety guidelines.

After probing with a radiolabelled antibody or protein A, the blot should be extensively washed under the most stringent conditions suitable for the given antigen/antibody system. In the case of high-affinity antibodies, the blot can be washed in the presence of 0.1% Triton X-100 plus 0.05% Tween-20 (Blotto-T, T, in *Protocol 7*) at room temperature for 10–20 h. If necessary, more stringent washing conditions such as high salt concentrations and various detergents (e.g. LiCl/NP-40 solution, *Protocol 7*) should be carefully explored.

Washed blots should be dried using paper towels, then placed on a cardboard support sheet of the same size as a film, and covered with a moisture-proof plastic wrap (e.g. Saran wrap). The wrap should be well stretched, smoothed down and fastened by tape at the back of the cardboard to prevent its wrinkling which could cause artefacts (creases or smudges) on the autoradiograph. A radioactive marker can be placed adjacent to the blot to facilitate orientation of the autoradiograph in relation to the blot. We made such markers using diluted solutions containing ^{32}P or ^{125}I, spotting about 500 c.p.m. in a domino-like pattern of pen dots on a little strip of filter paper. The strip was then taped over with Scotch tape and the whole marker taped on a cardboard sheet with a blot. Such markers, each with a unique dot pattern, are also very useful in preventing confusion when several films are developed together.

In most Western blotting experiments, the blot should be exposed to fast X-ray film (e.g. XAR-5 film from Kodak) with an intensifying screen, at -70 °C. The exposure time can be varied from several hours to 3 weeks as necessary, depending on the intensity of the specific signals and the level of background.

The molecular weight of protein bands on an autoradiograph of Western blot can be estimated by comparison with a molecular weight protein marker blotted along with the tested protein samples (*Section 8.2*). The mol. wt. marker, usually run in a side lane of an SDS-PAGE gel, can be either ^{125}I-labelled (use 100–500 c.p.m. per band) or prestained (the stained bands will be visible also on the blot). In the case of nitrocellulose blots, Amido black stained protein markers can be used instead (*Protocol 6*).

9.3 Re-used probes, reprobed blots, and multiple antibody probing

In general, diluted solutions of antibodies, protein A and protein G can be used several times as probes in Western blotting if the solutions are stored in a refrigerator and contain 0.02% merthiolate to prevent microbial growth. However, the stability of a probing solution should be examined for each particular reagent since some antibodies, mAbs in particular, lose specific antigen recognition even after a short storage. Destabilized probes usually bind additional, non-specific or cross-reactive protein bands when re-used (*Figure 9a*).

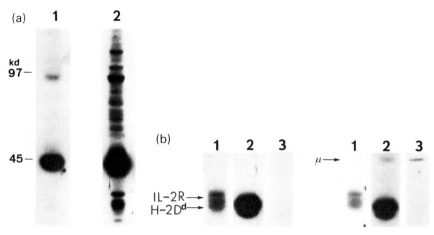

Figure 9. Recycled blots and reused probes in Western blotting. (a) Detection of the H-2D^d antigen using a fresh or an old, reused ^{125}I-labelled 7.2.14 mAb probe. Two samples of a Triton X-100 protein extract from BCL$_1$ B lymphoma cells (3×10^6 cell equivalents each) were resolved in a non-reducing 8% SDS-PAGE gel, and electroblotted on to a nitrocellulose filter. The H-2D^d antigen was detected by one-step probing with ^{125}I-labelled specific mAb 7.2.14 (31) (4×10^5 c.p.m./ml in Blotto-T). In **lane 1**, the 7.2.14 mAb probe was radioiodinated 3 days prior to using. For contrast, a 3-week-old ^{125}I-labelled 7.2.14 mAb probe, reused four times, was tested in **lane 2**. Both blot portions were washed overnight in Blotto-T, T and rinsed in the LiCl/NP-40 solution before autoradiography (20 h exposure). Note the increase in visualization of non-specific and cross-reactive protein bands associated with the use of the old, destabilized radiolabelled antibody probe in **lane 2**. (b) Detection of the H-2D^d antigen or the IL-2 receptor in activated B lymphocytes and subsequent detection of the IgM μ chain on the recycled blots. Purified splenic B cells were activated with 10 μg/ml LPS for 3 days and then lysed with 1% Triton X-100. Samples of the protein extract (3×10^6 cell equivalents each) were resolved in a reducing 8% SDS-PAGE gel and electroblotted on a nitrocellulose filter. The blot was cut into three strips (**lanes 1, 2** and **3**). The IL-2 receptor was detected by the specific rat mAb 7D4 (38) (p55, **lane 1**), the H-2D^d antigen by the specific rat mAb 7.2.14 (31) (p47, **lane 2**) while lane 3 was not incubated with any primary antibody. All three lanes were then treated with common secondary reagents: the mouse anti-rat kappa mAb RG7/9.1 (34) and ^{125}I-labelled protein A, and washed overnight in Blotto-T, T before autoradiography (26 h exposure). The blot strips were then re-probed: the IgM μ chain was detected in all three lanes by probing with a rabbit antiserum to the mouse μ chain followed by ^{125}I-labelled protein A. All the blot strips were washed and autoradiographed as before.

We have re-used probes containing different rat or murine mAbs, or rabbit antisera diluted in Blotto-T plus merthiolate for up to 2 weeks. In our experiments, diluted solutions of the ^{125}I-labelled antibodies or protein A appeared somewhat less stable than probes containing the undiluted reagents.

It is possible to detect two or more different antigens on one blot by subsequent (serial) probing with different primary antibodies. Limitations of this approach include (i) a complex band pattern after the secondary antigen detection because it is not possible to strip off the signal visualized by the first probing, and (ii) denaturation of the blotted antigens as well as an increase in fragility of

Detection of macromolecules

nitrocellulose membranes due to manipulations during the first probing and autoradiography. In our experiments, we have re-used nitrocellulose blots for serial detection of murine antigens by probing first with a specific mAb followed by detection with a polyclonal antibody (*Figure 9b*). In other experiments, a guinea-pig polyclonal antibody to LCM virus was used in the first probing and a mAb to class I H-2 antigens (20) was used subsequently. After autoradiography at $-70\ °C$, the blot was thawed, carefully removed from Saran wrap and reblocked by gentle agitation in Blotto-T at room temperature for 30 min. The blocked blot was then probed with the second specific primary antibody followed by standard secondary detection reagents. The sensitivity of antigen detection on a re-used blot seemed to be 3–5 times lower than on a fresh blot.

In an alternative approach, two or more different antigens can be visualized simultaneously by probing a blot with a solution containing two or more different primary antibodies. This parallel detection procedure provides the advantage of speed, similar sensitivity for the different antigens, and avoids protein denaturation and nitrocellulose damage caused by the serial re-use of a blot. However, parallel probing with several antibodies requires their compatibility in the subsequent detection technique. It is also important for the different antigens to have known and distinct relative molecular weights to allow their identification on the autoradiogram. The parallel multiantibody probing is potentially useful for first-round screening of an expresion library with a pool of mAbs followed by retesting of any positive clones with the individual mAbs.

10. RNA transfer and hybridization procedures

The origins of the current techniques for nucleic acid transfers to solid membrane supports can be identified amongst the classical methods for nucleic acid sequencing (reviewed in ref. 39). Since the first description of the DNA transfer process (40) many variations have been explored including the key finding that RNA, like DNA, can be transferred efficiently from an agarose gel to a nitrocellulose filter (41). The objective in the following sections is to provide details of RNA extraction, RNA gel electrophoresis, RNA transfer and hybridization procedures and to mention recent modifications to the well-established procedures that hold considerable promise for improved efficiency and convenience.

10.1 Extraction of total cell RNA by treatment with guanidinium thiocyanate (GTC)

10.1.1 Tissue culture cells

For cells in monolayers, discard the culture medium, rinse the cells once with PBS or normal saline and then add GTC (ref. 31, *Protocol 8*) directly to the flask (8.5 ml of GTC solution for approximately 2×10^8 cells in a T175 culture flask) and rock the solution back and forth over the monolayer for 2–3 min. The cells should lyse instantaneously. Leave the flasks horizontal with the GTC solution for 5 min then stand the flasks upright so that the thick, viscous solution drains to the bottom. Transfer the solution to a 50-ml screw-cap plastic tube and either shake the tube vigorously for 5 min or vortex with 3×30-sec bursts in order to shear the chromosomal DNA. RNAs up to 8–10 kb are not damaged by this extraction procedure. If necessary, the samples can be stored at this stage for up to 7 days at 4 °C although it is certainly preferable to proceed immediately with the CsCl gradients. For cells grown in suspension, the cells are pelleted, washed twice in a large volume of sterile PBS or saline (typically 20–40 ml/wash) and then resuspended in the appropriate volume of GTC for lysis ($1-2 \times 10^7$ cells/ml).

The RNA and, if required, the DNA can be recovered by layering the GTC cell lysate on top of a cushion of 5.7 M CsCl and centrifuging overnight. The RNA should form a tight pellet and the DNA will concentrate into a band in the upper half of the CsCl solution. Remove the GTC solution all the way down to the CsCl with a pipette and then, with a new pipette, remove the upper half of the CsCl. [This CsCl solution can be diluted 5-fold with 10 mM Tris–HCl, 1 mM EDTA pH 8.0 and precipitated with two volumes of ethanol to recover chromosomal DNA. Despite the physical treatment to shear the DNA prior to layering the GTC lysate on the gradients, the recovered DNA can be used for genomic DNA digests (hexanucleotide, pentanucleotide or tetranucleotide recognition sequences) to produce results that are very similar to those obtained with genomic DNA prepared in a more conventional manner.] Remove the remaining CsCl, taking care not to disturb the RNA pellet which, at this stage, should be a totally invisible 'glassy' disc located in the centre of the base of the centrifuge tube. The centrifuge tubes can be cut, approximately 1 cm above the base, in order to prevent a residue of DNA from sliding down the walls of the tube and contamination the RNA. In practice, handling the short tube stubs can create about as many problems as are intended to be avoided and rapid processing during the unloading of the gradients prevents anything other than a trace of DNA contamination. The RNA pellets are rinsed with 0.5 ml of cold 70% EtOH in 10 mM Tris–HCl pH 8.0, 1 mM EDTA, 50 mM NaCl and the RNA frequently detaches from the base of the tube to become visible as a semi-opaque floating form. For large amounts of RNA, this EtOH wash can be removed with a pipette and discarded; for small amounts of RNA (100 μg or less) this EtOH wash should be saved because part or all of the RNA will be transferred with the EtOH when it is removed. The RNA pellet is then resuspended in sterile water, transferred to a sterile tube and mixed with appropriate volumes of x10 TE and 5.0 M NaCl to reach final concentrations of x1 TE and 0.2 M NaCl and then precipitated by addition of 2–3 volumes of 95% EtOH. In this step, the RNA does not redissolve completely and we have found it more expeditious to transfer the partially hydrated RNA pellet than to produce a homogeneous RNA solution prior to the EtOH precipitation. If necessary, the RNA can be precipitated, redissolved and reprecipitated, prior to a final dissolving in sterile water. The RNA can be stored indefinitely either as an ethanol precipitate at -20 °C or as a homogeneous solution in sterile water at $-70\ °C$. For large RNA preparations and/or samples that will be used repeatedly, it is advantageous to divide the material into working amounts to avoid multiple freeze–thaw cycles with the entire sample.

RNA concentrations are determined by measuring the absorbance at 260 nm where 1 OD unit is equivalent to 40 μg/ml RNA. The ratio of absorbances for 260:280 nm should be approximately 2.0 (indicating an absence of protein) and the absorbance at 320 nm should <0.005 (indicating an absence of dust or other particulate matter in the sample). Deviations from these 280 and 320 nm readings indicate the presence of contaminants and the RNA should be extracted with phenol, precipitated from EtOH and redissolved.

Protocol 8. Guanidinium thiocyanate (GTC) extraction of RNA

Stock solution of GTC made up as follows:
- 4.0 M guanidinium thiocyanate–CAUTION: GLOVES!! GTC is a potent protein denaturant
- 25 mM Na citrate pH 7.0
- 0.5% Sarkosyl
- 0.1 M 2-mercaptoethanol
- 1–2 drops Sigma Anti-foam A

1. Dissolve the GTC solid in water—there is an enormous volume coefficient of expansion. Add solid Na Citrate and filter (Nalgene filter unit).
2. Add 10% Sarkosyl to 0.5%.
3. Adjust pH to 7.0 then add 2-mercaptoethanol and anti-foam.

The solution can be stored for 2–3 months at 4 °C in the dark.

4. Also needed: 5.7 M CsCl (Molecular Biology grade or better)
   0.1 M EDTA

Adjust pH to 7.0 (with conc. HCl).

5. Add DEP to 0.2%, leave overnight at room temperature, then store at 4 °C.
6. For cells in monolayers—discard the culture medium, rinse with PBS then discard PBS.
7. Add 8.5 ml of GTC per T175 flask; the monolayer should lyse instantaneously. Leave GTC on cells for 3–5 min then stand flasks upright to drain.
8. Remove the thick viscous solution and transfer to a screw-cap plastic tube. Shake vigorously to shear the DNA. Material can be stored in the cold at this stage although it is preferable to continue with the RNA extraction.
9. Polyallomer ultracentrifuge tubes must be used as GTC weakens the walls of polycarbonate tubes. Rinse with GTC solution to clean out the tubes, use a 2.0-ml cushion of CsCl per tube then overlay with the GTC soln. For the Beckman SW41 rotor do not use more than $\sim 2 \times 10^8$ cells (i.e. confluent T175 flask) per centrifuge tube.

Protocol 8. *Continued*

10. Spin overnight (≥ 12 h) at 15 °C, 30–35 k r.p.m. The RNA will pellet whereas the DNA will remain at the CsCl/GTC interface and in the upper portion of the CsCl solution. Note that this interface is not normally visible after the centrifugation.
11. Remove the GTC solution from above and wash the interface with 1.0 ml fresh GTC. Remove GTC and CsCl with a pipette.
12. Add 0.5 ml of cold 70% EtOH, ×1 TNE to the tubes and mix gently—you should see visible RNA precipitates forming.
13. With small amounts of RNA—remove and save this EtOH. With large amounts of RNA—remove and discard this EtOH.
14. Add 0.4 ml of sterile H_2O to the centrifuge tubes and transfer the partially redissolved RNA into sterile Eppendorf tubes.
15. Add stock solutions to obtain ×1 TE and 0.2 M NaCl final concentrations, fill the tubes with EtOH, mix and leave overnight at -20 °C to allow the RNA to precipitate.

10.1.2 Tissue samples

RNA can also be extracted by the GTC method from either fresh or frozen tissue with only minor variations in the basic procedure. For fresh tissue, mechanical homogenization in GTC produces very effective disruption. The time interval between animal sacrifice (or biopsy), tissue removal and disruption should be kept to an absolute minimum. Alternatively, the tissue can be removed, snap-frozen in liquid nitrogen and disrupted later. In this case, rather than homogenization it is probably advantageous to pulverize the frozen tissue and then simply place the powered tissue in GTC. A clearing spin (10 min at 8000 r.p.m.) is often helpful in removing clumps of debris, especially in the case of homogenization of fresh tissue, prior to layering the GTC lysate over CsCl cushions.

RNA from tissues is often not as pure as RNA from tissue culture cells after a single ultracentrifugation step. For analytical gels, the contaminants may not be problematic but the RNA could either be extracted with phenol or the centrifugation step could be repeated to obtain more highly purified RNA.

10.2 Phenol extraction

As an alternative to GTC disruption, RNA can be prepared by phenol extraction of either cells or tissues. A major advantage of this method is that cell lysis can be performed to allow separation of cytoplasm and nuclei and DNA contamination of the cytoplasmic RNA fraction can be largely eliminated.

Detection of macromolecules

10.3 Minigel analysis of RNA samples

As an initial check on the RNA extraction procedures, we have routinely analysed RNA samples on non-denaturing agarose minigels. This method allows visualization of host 28S and 18S ribosomal RNAs by ethidium bromide fluorescence. The ribosomal RNAs are relatively resistant to nuclease digestion and so, if the ribosomal RNAs do not appear as distinct bands it is likely that the sample has undergone extensive degradation. Such degradation probably would not affect the interpretation of dot blot or slot blot hybridization experiments but would prevent the derivation of meaningful information from RNA blotting or solution hybridization experiments. Additionally, the minigel analysis provides an independent verification of the RNA concentration and indicates the extent of DNA contamination. No special precautions are necessary for running the RNA minigels and a general method is summarized in *Protocol 9*.

Protocol 9. Electrophoresis of RNA on non-denaturing agarose minigels

1. Prepare a standard 1% agarose minigel in × 1 acetate buffer, exactly as would be used for analysis of DNA restriction digests. The agarose need not be autoclaved but it is preferable to prepare and use it immediately rather than remelting a stock preparation.
2. Add ethidium bromide (5 μg/ml) to the agarose, before pouring the gel, and also to the electrophoresis buffer. This eliminates the need for post-electrophoresis staining and allows the RNA bands to be visualized by illumination with UV light directly.
3. With whole cell RNA samples, 1 μg of RNA is sufficient to visualize host 28S and 18S ribosomal RNA species. Mix the necessary volumes of cell RNA samples with the appropriate volume of × 1 or × 10 buffer (either acetate electrophoresis buffer or restriction enzyme buffer) so that the RNA will be in × 1 buffer and the sample will fill the well.

If the gel is to be submerged in electrophoresis buffer, it is probably advantageous to use autoclaved buffer but for alternative electrophoresis arrangements this precaution is not normally required.

The 28S and 18S ribosomal RNAs should run on either side of the bromophenol blue marker dye and should appear as sharp, discrete bands. A rapidly migrating band at the base of the gel corresponds to 5S/5.8S size classes of RNA. Note that these low molecular weight RNAs are selectively under-represented when total cell RNA is purified by pelleting through a CsCl cushion (*Protocol 8*). If the ribosomal bands do not show up distinctly and/or there is intense ethidium fluoresence at the base of the gel then the RNA sample may have suffered extensive degradation.

11. Oligo (dT) cellulose chromatography

Cellular mRNAs can be separated from the vast excess of intracellular ribosomal RNAs by simple binding and elution from oligo (dT) cellulose. (Note that this oligo (dT) chromatography is also effective in separating mRNA from either viral or cellular DNA which otherwise might interfere with hybridization analysis.) The separation is based on the fact that almost all cellular (and viral) mRNAs have long (100–200 bases) 3' tails of poly (A) which can interact via hydrogen bonding with (dT) covalently attached to the cellulose. This interaction occurs in the presence of 0.5 M NaCl; non-polyadenylated RNAs do not bind and can readily be washed away. The mRNA is then eluted by washing the cellulose in sterile water. Optimally, a mRNA fraction should be passed over a second oligo (dT) cellulose column to ensure remove of most of the ribosomal RNA.

This separation can be performed using gravity flow of buffer through a small column although the columns can clog, which transforms a simple procedure into an extremely tedious exercise. The separation can also be performed effectively in a batch method to avoid slow flow rates in clogged columns. A good compromise is to perform the binding and washing batch-wise then to transfer the cellulose to a small column for the elution step.

Several RNA extraction kits are available from commercial sources that include a spin column method for oligo (dT) fractionation. One kit, from Invitrogen, San Diego, is particularly attractive because the RNA purification step has been eliminated and the oligo (dT) cellulose chromatography is performed directly with the cell or tissue lysate. This represents a significant saving in time, efort and materials. In our only test experiment to date, the Invirtrogen kit proved simple to use and came up to the published specifications.

12. RNA gel electrophoresis

There are three distinct methods in current usage for the electrophoretic separation of RNA molecules in agarose gels that involve denaturing the RNA with either methyl mercury hydroxide, formaldehyde or glyoxal. Each method confers specific advantages (and possibly some disadvantages) and, in general, there is a universal adoption of one method in any given laboratory.

Gels can be run either for short times at high current or for longer times (e.g. overnight) at low current with essentially the same quality of resolution. Just as with DNA gels, the percentage composition of agarose in the gel and the distance of migration can be manipulated according to the sizes of RNA species being examined. Typically, agarose concentrations in the range of 1–1.5% are used and, on a 20cm gel, it is possible to resolve RNA species in the approximate range of 400–15 000 bases. The methyl mercury or formaldehyde gels can be stained with ethidium bromide after the electrophoresis to visualize RNA bands within the gel although this may be somewhat detrimental to the subsequent transfer of RNA from the gel. An attractive alternative is to add ethidium bromide (1 μg) to the samples immediately before loading on to the gel so that the RNA bands can be visualized by UV irradiation as soon as the electrophoresis has been terminated. RNAs denatured by treatment with glyoxal are stained very inefficiently by ethidium bromide (both agents interact principally with G residues) and there is essentially no benefit from this step. In general, if it is necessary to verify that RNA species in the gel have electrophoresed appropriately, then the RNA bands can be visualized by UV shadowing (simply place the gel on top of a standard TLC plate covered with Saran wrap and illuminate from above with long wavelength UV light–RNA bands appear as shadows on the TLC plate). Problems arise either from degraded RNA and/or from RNA degradation during electrophoresis. If the original RNA sample has been shown to be intact when first analysed on a minigel (see *Protocol 9*) then RNA degradation is most probably occurring during sample preparation and/or electrophoresis.

A detailed procedure for the preparation of agarose gels and denaturation of RNAs with glyoxal and DMSO is given in *Protocol 10*. Early studies in the LCMV system had indicated that glyoxal treatment at 60 °C was necessary to achieve complete denaturation of viral RNAs and that the other denaturation systems were not as effective (43–45). Because we do not have extensive experience with the other gel systems we have not included practical information here and refer the reader to other excellent technical manuals (46).

Protocol 10. Agarose gel electrophoresis for RNA: glyoxal denaturation

- Stock buffer 1.0 M $NaPO_4$ pH 6.5 autoclaved. ($\times 100$)
- Agarose: made up in water (1% or 1.5%) and autoclaved. We use Seakem agarose (FMC)—other types of agarose may be equivalent. Heat in microwave to remelt.
- *NB*. Add requisite buffer ($\times 100$ stock) to agarose before pouring the gel.
- Apparatus must be clean and optimally is *only* used for RNA gel work. A reliable system for buffer recirculation is required otherwise there will be an enormous pH gradient created across the gel.
- RNA samples (from GTC or phenol preparations) are precipitated from EtOH, washed with 95% EtOH, dried and then redissolved in sterile H_2O. Determine RNA concentrations from spectrophotometer readings. Check RNAs on minigels for intactness (*Protocol 9*).
- Up to 100 μg of total cell RNA can be run on a 2-cm gel slot. Higher RNA concentrations will result in significant distortions in the regions of the host ribosomal RNAs.
- RNA samples should approximately fill the gel slots so adapt volumes to the following ratios of components

    RNA/H_2O 8.2 μl
    $\times$ 40 Stock buffer   1.0 μl
    (400 mM $NaPO_4$ pH 7.0, 10% SDS, 80 mM EDTA, autoclaved.)

Protocol 10. *Continued*

    Deionized glyoxal   3.8 μl

    (Stock glyoxal—40% reagent as purchased from Sigma or Kodak—deionized with mixed-bed resin, exactly as for formamide in hybridizations; store at −70 °C in small volumes. Stable indefinitely but do *not* refreeze.)

    DMSO   13.0 μl

    (Standard reagent—no special treatment.)

1. Mix at room temperature, vortex, spin and incubate for 3 min at 60 °C then for 15 min at 50 °C, then for ≥15 min at room temperature.
2. Add tracking dye (gel loading buffer: 10 mM NaPO$_4$ pH 7.0, 50% glycerol, bromophenol blue and xylene cyanol FF dyes; autoclave and store in the cold.) and load samples.

Electrophoresis buffer is 10 mM NaPO$_4$ pH 6.5 and it is not normally necessary to use sterile buffer unless the gel is to be submerged in buffer throughout the duration of the electrophoresis.

3. Prerun the gel for 15–20 min with wells filled with 10 mM NaPO$_4$ buffer plus 0.1% SDS—the SDS will run through the gel ahead of the RNAs.

13. Preparation of radioactively labelled hybridization probes

13.1 Synthesis of double-stranded DNA probes

As a standard procedure, we have decided to use purified restriction fragments rather than intact plasmid or phage DNAs as the starting material for DNA labelling reactions. This does require the additional steps of fragment purification but there is the substantial benefit that all of the incorporated label is specific to the insert sequence rather than being randomly distributed throughout the genome of the recombinant vector. We have used either electroeution into dialysis tubing or the NaI/glass bead technique with high efficiency and reproducibility to recover restriction fragments from agarose gels. DNA eluted from glass beads should be precipitated by addition of NaCl, to a final concentration of 0.1 M and 3 volumes of 95% EtOH. This will remove traces of salt (principally NaI) that can be carried forward through the washing steps and which would otherwise produce a dramatic inhibition of restriction enzymes, DNA ligase or Klenow DNA polymerase. Fragments longer than 250 bp can readily be labelled using the hexanucleotide primer method (47), resulting in incorporation of 50% or more of the input labelled nucleotide. A potential disadvantage of this method is the difficulty in determining the specific activity of the labelled DNA product. However, it is possible to express total counts incorporated as a function of input DNA and, we have observed impressive reproducibility when using the same amount of the same fragment in a reaction that is assembled according to a standard protocol (*Protocol 11*).

Protocol 11. Labelling of DNA with random hexanucleotide primers

Stock Solutions A
- Hexanucleotide primer pd(N)$_6$
- Pharmacia #27-2166-01
- Dissolve at 90 OD units/ml in x1 TE
- Store at $-20\,°C$

Stock Solutions B
- Sterile 1.0 M Hepes pH 6.5

Stock Solutions C
- 250 mM Tris–HCl pH 8.0
- 25 mM MgCl$_2$
- 50 mM DTT
- 100 μM dGTP
- 100 μM dTTP

    Keep 10 mM stocks of dNTPs—do not store or re-use buffer C.

    1. Mix solutions A:B:C in ratios 7:25:25 = LS buffer.

    2. Assemble reactions in the presence of 50 μCi of dried [α-^{32}P]dATP and 50 μCi [α-^{32}P]dCTP. Final volume 25.0 μl:

    11.4 μl   LS buffer
    1 μl   10 mg/ml BSA
    12 μl   DNA/H$_2$O (see below)
    0.6 μl   Klenow DNA polymerase ($\sim$1–3 units)

- Ideally use 50–100 ng of DNA per reaction. The DNA/H$_2$O mix in a final volume of 12 μl should be heated for 5 min in a boiling water bath and chilled on ice immediately prior to addition to the reaction. Incubate the reaction mix at room temperature overnight or for 2–3 h at 37 °C. Stop the reaction with excess EDTA and purify the labelled DNA using conventional column chromatography or spin columns.

- Normally we will purify specific restriction fragments for labelling rather than labelling plasmid plus insert. This is slightly more work but then 100% of the counts will correspond to the insert sequence.

- Probes prepared by this method will approach 10^9 c.p.m./μg input DNA and should be used as soon as possible after synthesis. Probes more than one week old may show significant reduction in signal and/or increased non-specific binding as a consequence of radionuclide damage.

In a typical labelling reaction with 50–100 ng of input DNA, we expect to recover 0.2–1.0×10^8 c.p.m. of labelled product that will have a useful life of up to two weeks when stored at 4 °C. With extended storage, significant radionuclide damage can occur and the resultant fragmentation of the probe can cause non-specific background problems. Because of this problem and the relatively short half-life (14.2 days) of ^{32}P, probes should be used within a few days of synthesis.

13.2 Synthesis of single-stranded probes

Two separate methods have been used in the recent past for the synthesis of uniformly labelled single-strand probes. Both methods require the use of particular vector systems (M13, λ or plasmids) and necessitate either a subcloning step to transfer the target sequences or primary cloning into this type of vector. Consistent background problems have been observed if G–C sequences (arising from cDNA cloning into *Pst*I sites by G–C tailing) are present in these vectors and would therefore be included in the single-stranded probes. A subcloning strategy that would eliminate G–C tails is strongly recommended. A DNA strand can be synthesized from a M13 template or RNA can be synthesized *in vitro* using a cloned SP6, T7 or T3 prokaryotic promoter in the presence of the appropriate purified RNA polymerase. Both methods yield high quality, high specific activity probes but there is an emerging preference for the RNA systems. The relevant technical information has been reviewed elsewhere in this series (48).

DNA templates destined for *in vitro* RNA probe synthesis need to be cleaved with a restriction enzyme at the 3' side of the transcribed region. This creates a precise 3' terminus for the RNA product because the polymerase molecules fall off the end of the template. The cleaved DNA can be prepared in large amounts and, after enzyme digestion, analytical gel electrophoresis, phenol extraction and EtOH precipitation, can be stored indefinitely at −20 °C.

13.3 Gel transfer procedures

A major advantage with glyoxal denaturation and electrophoresis on neutral agarose gels is achieved at the stage of RNA transfer. With glyoxal gels, the RNA can be transferred directly without any pretreatments. The capillary transfer method probably remains the current method of choice but the availability of inexpensive and efficient vacuum blotting devices is likely to lead to a revision of the standard protocol. In the capillary transfer method, buffer (normally ×20 SSC) is drawn through the gel and, in the process, RNA is eluted from the gel and immediately comes into contact with a nylon or nitrocellulose membrane. Typically, the transfer is left for several hours at room temperature and it is often convenient to leave this transfer to occur overnight. The filters must then be baked (80 °C, 1–2 h in a vacuum oven for nitrocellulose, or 80 °C, 1–2 h for nylon—the vacuum oven is not strictly required for nylon membranes but it is by no means disadvantageous) or irradiated (UV Stratalinker, Stratagene Cloning Systems, San Diego, CA) to cause essentially irreversible binding of the RNA to the membrane. The UV crosslinking is rapidly becoming established in a

Detection of macromolecules

streamlined blotting procedure and the Stratalinker is an important new technical asset.

13.4 Hybridization and washing conditions

Details for prehybridization, hybridization and washing of filters are presented in *Protocols 12* and *13*. Minor variations may be advantageous depending upon the type of membrane used and the type and amount of probe but such variations can often only be recognized on an empirical basis. Many problems can be explained in the following categories:
- inadequate/incomplete washing after hybridization.
- excess probe and/or degraded probe.
- probe contaminated with unincorporated nucleotides.
- incomplete mixing during prehybridization or hybridization.

Protocol 12. Prehybridization and hybridization conditions

Final concentrations
- ×5 SSC
- ×2.5 Denhardt's solution
- 50 mM $NaPO_4$ pH 6.5
- 0.05% SDS
- 0.1% Sarkosyl
- 50% Deionized formamide
- 200 µg/ml Boiled, sonicated carrier DNA

Keep ×20 SSC, ×10 Denhardt's solution as a stock at $-20\,°C$

We have found this solution to be suitable for prehybridization and hybridization of both nitrocellulose and nylon (Biotrans, ICN) membranes using either ^{32}P or digoxigenin-labelled probes. Typically, the prehybridization period is for 2–4 h or overnight at 37 °C or 42 °C using 10 ml of solution per 200 cm^2 of membrane. The hybridization reaction is performed with a fresh preparation of the same solution, typically 4 ml of solution per 200 cm^2 of membrane, for 24–36 h at 37 °C or 42 °C. Minor adjustments to this mix may be necessary, in the light of manufacturer's suggestions, for other types of nylon membranes.

Protocol 13. Washing conditions

1. Remove hybridization solution from heat-seal bag and either store for reuse or discard as ^{32}P liquid waste.
2. Rinse filter in ~100 ml of ×2 SSC, 0.1% SDS and also discard this solution as ^{32}P liquid waste.
3. Wash twice in 300 ml per wash of ×2 SSC, 0.1% SDS at 37 °C. 20–30 min per wash.
4. Wash once in 300 ml of ×2 SSC, 0.1% SDS at 55 °C. 20–30 min.

Protocol 13. *Continued*

5. Rinse the filter twice and then wash once in 300 ml of × 0.1 SSC, 0.1% SDS, 0.1% Tween-20 at 55 °C. 30 min, timed carefully for this final wash.
6. Remove the filter from the final washing solution, allow the excess liquid to drain away then wrap the damp filter in Saran wrap and expose for autoradiography.

13.5 Reuse of hybridization filters

Nitrocellulose or nylon membranes can be recycled with different probes once suitable autoradiographic exposures have been obtained with the first probe. In this regard, the nylon membranes are much superior to conventional nitrocellulose because, even with the most careful handling, most nitrocellulose filters do not survive for more than four or five cycles whereas the nylon membranes are essentially indestructible. If probe removal and reuse are projected then the filters should not be allowed to dry out at any stage during autoradiography or storage prior to recycling. Probes can be stripped by washing the filters for 1–2 h at 90 °C in 0.1 × SSC, 0.1% SDS, 0.1% Tween-20 or by pouring boiling water on to the filter and leaving it to cool to about 60 °C (process repeated twice). For critical experiments, it is advisable to expose the stripped filter to X-ray film for a realistic time period (i.e. related to the anticipated time required to see a positive signal in the experiment) to verify that all of the previous signal has been removed. After stripping, the prehybridization step should be repeated prior to hybridization with a new probe. In this manner, considerable information can be accumulated from a single filter and the relative hybridization signals can be compared because the individual probe signals all relate to the same starting sample. Nucleic acid does leach off the filters at a very slow rate but, for most purposes, this loss can be regarded as insignificant. If totally rigid comparisons are required, then in the final hybridization reaction, the filter can be treated again with the very first probe so that matched exposures with equivalent probes of the same sequence specificity could be used to estimate the loss of target nucleic acid from the filter.

14. Recent technical developments

We have begun to evaluate two newly developed products—a vacuum blotting device for nucleic acid transfer and a non-isotopic hybridization detection system—that provide alternatives to widely accepted and established methods. Both new methods have attractive theoretical advantages over the conventional methods and the initial results in our laboratory have been very impressive. Clearly it will require a substanial investment of time to establish total compatibility with all experimental requirements but it is already apparent that standard analyses can be performed efficiently and reproducibly with substantial savings in time and effort using the new methods.

14.1 Vacuum blotting

There is an almost universal acceptance of the method of capillary diffusion of buffer (normally × 10 or × 20 SSC) to transfer nucleic acids from gels to nylon or nitrocellulose membranes (40). The method is simple and reliable but does require a time interval of several hours to ensure complete transfer of nucleic acids from the gel and it is often convenient to leave the transfer in progress overnight. A method for direct hybridization to nucleic acids immobilized within dried agarose gels has been available for several years (49) but has not apparently gained widespread popularity. The vacuum blotting approach involves a combination of these two previous methods as the nucleic acid within the gel is drawn on to the hybridization membrane by application of gentle suction to the gel. Small volumes of transfer buffer (again, × 10 or × 20 SSC) are required to keep the gel wet during the vacuum blotting. The transfer is essentially complete within 45–60 min and then the filter is processed according to existing protocols.

We have first-hand experience with the transvac vacuum blotter (TE80, Hoefer Scientific Instruments) and an adjustable vacuum pump (PV100, also from Hoefer) but similar units are available from several other sources. The adjustable vacuum pump appears to be crucial to the success of the method as a strong vacuum will cause the matrix of the gel to collapse thereby trapping nucleic acid within the gel. To date, we have demonstrated that DNA fragments can be transferred efficiently to either nitrocellulose (Schliecher and Schuell, 0.45 μm) or nylon (Biotrans Nylon, ICN, 0.2 μm) membranes after completion of standard procedures for depurination, denaturation and neutralization. Essentially all the higher molecular weight DNA from a genomic DNA digest could be removed from the gel in a 45–60 min vacuum transfer and none of the DNA could be detected passing through the membrane. For RNA gels (1–1.5% agarose in 10 mM $NaPO_4$ pH 6.5 for glyoxal treated RNAs), we have found that nylon membranes (Biotrans, ICN) must be used for quantitative binding of the RNA as it is drawn out of the gel. With nitrocellulose filters (Schleicher and Schuell, 0.45 μm), a substantial fraction of the transferred RNA passed through the filter and could be trapped on a second (nylon) filter. We are continuing to evaluate the general applicability of the vacuum transfer method for DNA and RNA molecules of differing lengths because, for example, after depurination, DNA fragments in the size range 0.5–1.0 kb are retained much less efficiently than DNA fragments of 5 kb or larger.

14.2 Non-isotopic hybridization probe strategies

There are several advantages that could be conferred by non-isotopically labelled hybridization probes over conventional ^{32}P, ^{35}S- or ^{125}I-labelled probes if equivalent sensitivity and specificity could be achieved. The non-isotopic probes would represent a convenience in the research laboratory but will probably be essential if nucleic acid hybridization technology is to be introduced effectively into the clinical diagnostic laboratory. Briefly, the principal benefits of non-isotopic probes can be summarized:

- Large-scale preparation and long-term storage of probes to facilitate reproducibility in hybridization assays.
- Elimination of radiation exposure hazards to laboratory workers and avoidance of problems with the disposal of radioactive waste.

The original non-isotopic hybridization probe schemes were based on a biotin derivative nucleotide (dUTP or, more recently, dATP) in which the biotin was attached to the base via a spacer arm. The biotin derivative nucleotides can be incorporated into DNA by Klenow DNA polymerase in either the standard nick translation or hexanucleotide primed labelling reactions. Biotinylated DNA is detected either with an antibody directed against biotin or in sandwich amplification systems using streptavidin/avidin and a biotinylated antibody. In both cases, these antibodies are conjugated to enzymes (most commonly, alkaline phosphatase or horseradish peroxidase) and, in the presence of the appropriate substrate, an insoluble, coloured product is precipitated where probe–target duplex structures have formed.

We have recently initiated a systematic evaluation of a new system that is based on the conjugation of the hapten, digoxigenin, via a spacer arm to dUTP and an F(ab) antibody specific for digoxigenin that is conjugated to alkaline phosphatase [both available from Boehringer Mannheim either as individual reagents or as components of the Genius labelling kit. (50)]. In dot blot, slot blot, genomic DNA transfers, and RNA transfers we have observed quite impressive sensitivity with these non-isotopic probes and, in a highly illustrative but qualitative comparison, the limit of detection with the digoxigenin system is approximately equivalent to a 48–72-h exposure for a ^{32}P probe at $-70\,°C$ with an intensification screen.

15. Ribonuclease contamination

There is no doubt that the technical demands for successful RNA experimentation are substantially greater than for similar DNA work. However, careful attention to several key points (*Protocol 14*) should be sufficient to lay the ribonuclease spectre to rest. Many laboratories insist on oven-baked glassware for all RNA manipulations but, in our experience, tissue culture grade plasticware can serve as a readily available substitute without jeopardizing the experimental results.

Protocol 14. Essential requirements for trouble-free RNA experimentation

- Reliable source of distilled or deionized and filtered water. Trace metal ion contaminants can create a major problem.

Protocol 14. *Continued*

- Autoclaved solutions, all made using the reliable water source and autoclaved tips and tubes.
- Wear gloves at all times.
- If possible, retain a separate electrophoresis apparatus exclusively for RNA work.
- Avoid repeated freeze–thaw cycles with stock RNAs and store all stocks at $-70\ °C$ in sterile water.

References

12. Towbin, H., Staehelin, T., and Gordon, J. (1979). *Proc. Natl. Acad. Sci. USA*, **76**, 4350.
13. Towbin, H., Schoenenberger, C., Ball, R., Braun, D. G., and Rosenfelder, G. (1984). *J. Immunol. Methods*, **72**, 313.
14. Talbot, P. J., Knobler, R. L., and Buchmeier, M. J. (1984). *J. Immunol. Methods*, **73**, 177.
15. Bers, G. and Garfin, D. (1985). *Biotechniques*, **3**, 276.
16. Hawkes, R., Niday, E., and Gordon, J. (1982). *Anal. Biochem.*, **119**, 142.
17. Lipkin, W. I. and Oldstone, M. B. A. (1986). *J. Neuroimmunol.*, **11**, 251.
18. Knudsen, K. A. (1985). *Anal. Biochem.*, **147**, 285.
19. Fernandez-Pol, J. A., Klos, D. J., and Hamilton, P. D. (1982). *Biochem. Intl.*, **5**, 213.
20. Haeuptle, M. T., Aubert, M. L., Djiane, J., and Kraehenbuhl, J. P. (1983). *J. Biol. Chem.*, **258**, 305.
21. McLellan, T. and Ramshaw, J. A. M. (1981). *Biochem. Genet.*, **19**, 648.
22. Hayman, E. G., Engvall, E., Hearn, E. A., Barnes, D., Pierschbacher, M., and Ruoslahti, E. (1982). *J. Cell. Biol.*, **95**, 20.
23. Co, M. S., Gaulton, G. N., Fields, B. N., and Green, M. I. (1985). *Proc. Natl. Acad. Sci. USA*, **82**, 1494.
24. Bowen, B., Steinberg, J., Laemmli, U. K., and Weintraub, H. (1980). *Nucleic Acids Res.*, **8**, 1.
25. Hendrickson, W. (1985). *Biotechniques*, **6–7**, 198.
26. Womack, M. D., Kendall, D. A., and MacDonald, R. C. (1983). *Biochim. Biophys. Acta*, **73**, 210.
27. Johnson, D. A., Gautsch, J. W., Sportsman, J. R., and Elder, J. H. (1984). *Gene Anal. Tech.*, **1**, 3.
28. Harlow, E. and Lane, D. (1988). *Antibodies, A laboratory manual*. Cold Spring Harbor Laboratory Press, NY.
29. Lindmark, R., Thoren-Tolling, K., and Sjoquist, J. (1983). *J. Immunol. Methods*, **62**, 1.
30. Ziff, M., Brown, P., Lospalluto, J., Badin, J., and McEwen, C. (1956). *Am. J. Med.*, **20**, 500.
31. Southern, S. O., Swain, S. L., and Dutton, R. W. (1987). *J. Immunol.*, **138**, 2568.
32. Southern, S. O., Swain, S. L., and Dutton, R. W. (1989). *J. Immunol.*, **142**, 336.
33. Southern, S. O. and Dutton, R. W. (1989). *J. Immunol.*, **142**, 3384.
34. Springer, T., Bhattacharya, A., Cardoza, J., and Sanchez-Madrid, F. (1982). *Hybridoma*, **1**, 257.

35. Bordier, C. (1981). *J. Biol. Chem.*, **256**, 1604.
36. Radka, S. F., Machamer, C. E., and Cresswell, P. (1984). *Human Immunol.*, **10**, 177.
37. Kaetzel, C. S., Mather, H., Bruder, G., and Madara, P. J. (1984). *Biochem. J.*, **219**, 917.
38. Malek, T. R., Robb, R. J., and Shevach, E. M. (1983). *Proc. Natl. Acad. Sci. USA*, **80**, 5694.
39. Murray, K. and Old, R. W. (1974). *Progress in Nucleic Acids Research and Molecular Biology*, Vol. 14, p. 117.
40. Southern, E. M. (1975). *J. Mol. Biol.*, **98**, 503.
41. Thomas, P. S. (1980). *Proc. Natl. Acad. Sci. USA* **77**, 5201.
42. Chirgwin, J. M., Przybyla, A. E., MacDonald, R. J., and Rutter, W. J. (1979). *Biochemistry*, **18**, 5294.
43. McMaster, G. K. and Carmichael, G. C. (1977). *Proc. Natl. Acad. Sci. USA* **74**, 4835.
44. Dutko, F. J., Kennedy, S. I. T., and Oldstone, M. B. A. (1981). In *The Replication of Negative Strand Viruses*, (Ed. D. H. L. Bishop and R. W. Compans), p. 43. Elsevier, Amsterdam.
45. Southern, P. J., Singh, M. K., Riviere, Y., Jacoby, D. R., Buchmeier, M. J., and Oldstone, M. B. A. (1987). *Virology*, **157**, 145.
46. Ogden, R. C. and Adams, D. A. (1987). *Meth. Enzymol.*, **152**, 61.
47. Feinberg, A. and Vogelstein, B. (1983). *Anal. Biochem.*, **132**, 6.
48. Little, P. F. R. and Jackson, I. J. (1987). In *DNA Cloning: A Practical Approach*, Vol. III. (ed. D. M. Glover), p. 1. IRL Press, Oxford.
49. Shinnick, T. M., Lund, E., Smithies, O., and Blattner, F. R. (1975). *Nucleic Acids Res.*, **2**, 1911.
50. BMBiochemica Volume 5, #4, July 1988.

2C. NUCLEIC ACID EXTRACTION AND DETECTION FROM PARAFFIN-EMBEDDED TISSUES

MATTHIAS LÖHR and MICHAEL I. NERENBERG

16. Introduction

Direct analysis of viral associated nucleic acids is a powerful technique for diagnosing viral infections in man (see Chapter 1). Ideally, one would work with fresh or frozen samples. However, limited access makes this impractical in many cases. On the other hand, a vast, under-utilized resource exists in most departments of pathology where libraries of paraffin-embedded tissue specimens are stored. A wide spectrum of samples are available from patients which have been extensively characterized by serological and clinical criteria. In addition, the biopsies have been categorized morphologically.

A variety of techniques may be used to detect viruses. Immunocytochemistry is simple, quick and allows cellular localization but requires access to well characterized antibodies which are able to bind antigens in formalin-fixed and paraffin-embedded specimens. In addition, persistent or latent infections resulting in small amounts of early viral proteins may not be detected. *In situ* RNA hybridization analysis (Chapter 3.3) is another method of detecting and localizing viral nucleic acids, however, it is rather insensitive, requiring 10–100 copies/cell (Chapter 3.2).

Recently, techniques have become available for purifying and concentrating nucleic acids from paraffin-embedded specimens (51, 52). Wax impregnation stabilizes nucleic acids and limits degradation even if stored for decades (53).

The advantages of this approach are:

(a) The technique is simple and requires no special equipment other than a microtome.
(b) Viral nucleic acids can be concentrated and detected by sensitive filter hybridization techniques.
(c) Probes for specific viral genes may be easily derived using published sequence data.
(d) Paraffin blocks are not harmed by these methods and are available for subsequent *in situ* analysis.

The disadvantages are:

(a) Loss of anatomical localization.
(b) Recovery of partially degraded nucleic acids.

The latter problem can be partially overcome by use of sensitive detection methods.

17. General considerations

Though widely available, paraffin-embedded tissues are frequently not ideally processed for the preservation of nucleic acids, especially RNA. Use of surgical pathology specimens is preferable. However, this is usually not possible for a variety of organs (e.g. brain, pancreas). Therefore, post mortem specimens are frequently used for analysis. Since fixation inactivates nucleases, the most crucial factor is the rapidity with which live tissue is fixed. Delays may occur in removal of tissue (as frequently occurs in post mortem biopsies), or from delays in permeation of fixatives (lung biopsies, etc.). These factors are most crucial in tissues with high levels of endogenous nucleases [e.g. pancreas (5)]. Further, the kind of fixative (51, 52), length of fixation, paraffin embedding and storage can potentially influence the preservation of nucleic acids. The ideal conditions would be as follows:

(a) shortest possible time lag from biopsy;
(b) immersion in fixative (10% buffered formalin), 24-h fixation;
(c) direct processing into paraffin.

Generally, these factors are not known and cannot be influenced within a retrospective study. One should therefore anticipate varying degrees of degradation of nucleic acids. We therefore routinely perform analysis of extracted nucleic acids by agarose gel electrophoresis. Nucleic acids stained with ethidium bromide are compared against known size DNA or RNA markers, in order to check for the quality of our preparations. Samples with extensive degradation are not eligible for further analysis and should be excluded (see below). DNA is less susceptible to degradation and therefore fewer samples need be excluded.

Isolation of nucleic acids from paraffin-embedded tissue is done in three steps: firstly, the wax has to be removed completely and tissue brought into an aqueous environment. Secondly, the nucleic acids, trapped and crosslinked in a mesh of protein, have to be released with minimal degradation. This is achieved by digesting away the protein in the presence of a high concentration of SDS, a chaotropic agent, which facilitates penetration and simultaneously inactivates nucleases. Proteinase K digestion is performed in 2% SDS at elevated temperatures. Competitive nuclease inhibitors such as RNAsin are inactivated under such conditions and are therefore not included. Finally, nucleic acids must be purified from other cell components.

18. Sectioning paraffin blocks

Prior to using the tissue for extraction of nucleic acids, at least one section should be cut and stained with haematoxylin and eosin. Blocks with extensive necrosis or poor preservation should be eliminated. Paraffin blocks should be precooled at $-20\ °C$ prior to analysis.

Sectioning of tissue, the first step of disrupting the cellular architecture, is done with a standard microtome. Usually 30 5-μm thick sections are cut from normal size biopsies, yielding about 100–200 mg of tissue. Smaller specimens require more sections. Paraffin ribbons are collected into preweighed 15 cm^3 polypropylene tubes and the ratio of tissue to paraffin is noted. After collecting tissue, the tubes are reweighed and the amount of tissue present can be calculated. Sequential dewaxing, hydration, and digestion steps may be performed in polypropylene tubes. Alternatively, spin columns (Bio-Rad) can be used. They facilitate solvent exchange without tissue loss and the tops and bottoms may be capped during the digestion. Tissue is trapped on the nitrocellulose frit and solvents may be dripped or spun through.

19. Extraction of RNA and DNA

Although the method for the extraction of RNA and DNA is similar, there are important differences. The different buffers for processing the tissue are shown in *Table 2*. The size of RNA is comparatively small in relation to DNA. Therefore, it is more readily released from tissue. On the other hand, RNA is unstable at high

Table 2. Digestion buffer for DNA and RNA

DNA	RNA
500 mM Tris–HCl pH 9.0	500 mM Tris–HCl pH 7.4
10 mM EDTA	20 mM EDTA
10 mM NaCl	10 mM NaCl
1% SDS	1% SDS

pH and more susceptible to chemical degradation and ubiquitous RNAses. RNA digestion is therefore performed at a lower pH and for a maximum of 48 h. Eighteen hours is usually sufficient and prolonged digestion is counterproductive. On the contrary, release of large molecular weight genomic DNA requires considerably more digestion time. Digestion is performed at high pH to facilitate degradation of RNA. Digestion extending from 7 to 10 days is usually necessary.

(a) Up to 200 mg of tissue is resuspended in 1 ml of digestion buffer (see *Table 2*) with vigorous vortexing.

(b) Proteinase K is added to a final concentration of 500 μg/ml (51, 52). After gentle mixing, incubation is performed at 37–55 °C (typically 48 °C) for 24 h.

(c) After vortexing, the mixture is supplemented with additional SDS and proteinase K to a final concentration of 2% and 1 mg/ml, respectively. For RNA, additional incubation is 6–12 h (57), for DNA 7–10 days (52).

(d) Optimal yield of nucleic acid depends on the relative amount of connective tissue collagen (52) and the degree of crosslinking. In general, it takes at least 5 days to release large molecular weight DNA.

After digestion is completed, DNA samples are extracted twice with equal volumes of phenol/chloroform (1:1 ratio) and once with chloroform. The final NaCl concentration is adjusted to 500 mM (300 mM NaOAC may be substituted) and two volumes of ethanol are added. Samples may be chilled overnight at -20 °C or rapidly frozen in a dry ice ethanol bath. Samples are precipitated by centrifugation in the cold. Fifteen minutes in a microcentrifuge set at high speed (10 000 g) is sufficient. The DNA is resuspended in TE buffer (10 mM Tris–HCl pH 7.5 and 1 mM EDTA) (usually 50 μl) and an aliquot quantitated by spectroscopy. One to five micrograms are checked by agarose gel electrophoresis for quality and contamination. A 0.8% agarose gel containing 0.5 μg per cm^3 ethidium bromide is run in TAE buffer (0.04 M Tris, pH 8.0, 0.002 M EDTA, 0.906 M acetate). Proteinase K digested RNA samples are diluted with 8 ml of 4 M guanidium isothiocyanate (GTC) and either extracted with hot phenol followed by ethanol precipitation (8) or pelleted through a cushion of caesium chloride by overnight centrifugation (58). Both methods are explained in *Protocol 15*. Though the yield of RNA is considerably less when pelleted through a caesium gradient, DNA contamination is greatly diminished.

Protocol 15. RNA extraction methods

Guanidium/hot phenol method (8)
1. Diulte 1 ml of digest with 4 ml of 4 M GTC.
2. Bring the mixture to 60 °C and force through a syringe with an 18 G needle.
3. Add an equal volume to phenol, preheated to 60 °C, and force the mixture again through the syringe.
4. Add 0.5 vol of (0.1 M NaOAc, pH 5.2; 10 mM Tris–HCl, pH 7.4; 1 mM EDTA).
5. Add 1 equal volume of a 24:1 chloroform/isoamyi/alcohol mixture.
6. Shake vigorously for 10–15 min at 60 °C.
7. Cool on ice, centrifuge at 4 °C and 2000 g for 10 min.
8. Recover the aqueous phase, re-extract with phenol/chloroform. Centrifuge.
9. Re-extract the aqueous phase with chloroform.
10. Add 2 vol of cold ethanol (100%). Store at -20 °C overnight.
11. Recover the RNA by centrifugation at 4 °C and 12 000 g for 20 min.
12. Rinse the pellet in 70% ethanol. Dry.
13. Resuspend the pellet in DEPC treated (0.1%) sterile water.

Guanidium/caesium chloride method (8)
1. Add 1 ml of digest to 8 ml of 5 M GTC.
2. Layer carefully over 2 ml of caesium chloride (5.7 M in 0.1 M EDTA) in a Beckman SW 40.1 polyallomer tube (rinsed with GTC).
3. Centrifuge at 35 000 r.p.m. and 20 °C overnight.
4. Suck off GTC, rinse with fresh GTC.
5. Suck off GTC and caesium chloride up to $\simeq 1$ ml left.
6. Invert the tube and drain $CsCl_2$ dry wall of the tube.
7. Resuspend the pellet in TE with 1% SDS.
8. Extract twice with an equal volume of phenol.
9. Extract twice with an equal volume of chloroform.
10. Add 0.1 vol of 3 M NaAc (pH 5.2) and 2.2 vol of cold ethanol (100%).
11. Store at -20 °C overnight.
12. Then allow the same procedure as under *A*, steps 11–13.

A comparison of different methods are given for pancreatic tissue in *Table 3*. Fixation in 10% buffered formalin gives the highest yields though we have also succeeded in isolating RNA from samples fixed in Bouin's solution. The latter required repeated ethanol washes to clear the picric acid (see below) but let to

Table 3. Different extraction protocols for RNA from pancreas

Tissue	Procedure	Tissue (mg)	RNA (mg)	Relative yield
Fresh	GTC/CsCl$_2$, ultra centrifugation; phenol/chloroform[a]	1000	12.0	100
Formalin	GTC/CsCl$_2$, ultra centrifugation; phenol/chloroform	2400	0.856	3
	GTC; proteinase K; hot phenol/chloroform	1150	2.781	20
Paraffin	Proteinase K; GTC/CsC1$_2$, ultra centrifugation; phenol/chloroform	195	0.027	1.1
	Proteinase K; GTC; hot phenol/chloroform	234	0.730	26

[a] GTC = guanidium thiocyanate

reasonable amounts and quality of RNA. Tissue fixed in Zenker's solution is reported to be not suitable for isolation of nucleic acids, and has not been used in our experiments.

The yield of DNA and RNA can be calculated and varies between tissues. For DNA, this roughly correlates with the nuclear to cytoplasmic ratio of constituent cells. This reflects the relative amount of nuclei per volume of tissue. For example, lymphatic tissue will yield comparatively more DNA than brain or connective tissue.

20. Analysis of nucleic acids

Nucleic acids may conveniently be analysed by dot or slot blot hybridization (50, 51, 53, 54, 58) or by polymerase chain reaction (PCR) (52, 59–63).

By the nature of the starting material, there will be degradation of nucleic acids to a variable extent. Blotting on standard nitrocellulose filters will trap many of these degraded products which are suboptimal for hybridization and will diminish the signal. Hybridization of slot blots with probes to endogenous genes (e.g. 28S ribosomal, or β-actin) can serve as internal controls and are routinely used in our studies. The signal with the indicator probe partially corrects for the variation arising from degraded RNA/DNA rather than a true difference in the signal. Negative controls should include tissue processed the same way, that has been fixed and paraffin-embedded. It is also useful to include dose titrations of a positive control. A convenient way to do this is to dilute plasmid containing the gene of interest in negative control DNA. We routinely include 1/10 1, and 10 copy per cell equivalents. Southern or Northern blotting allows additional identification of the positive signal by predicted size, but in many cases is not possible because of degradation. Signals may be quantitated by laser densitometry (LKB or other models) (64).

Polymerase chain reaction analysis extends the sensitivity of detection by at least three orders of magnitude (see Chapter 3.5). By adjusting conditions to amplify only the gene of interest the level of signal to noise is greatly reduced. In a typical 30-cycle reaction, viral sp. sequences may be amplified by 10^5 or more. A typical DNA amplification will span a DNA stretch between 130 and 500 BP. The amplified DNA in turn can be transferred to nitrocellulose by Southern blot and hybridized to specific probes. In some instances, genes or portions of genes may be cloned, either after restriction enzyme digest of the isolated DNA (52) or from the PCR product (Chapter 3.5).

If there are any breaks in the DNA sequence between the oligonucleotide primers, the PCR may still yield the expected size product by jumping to other fragments. Alternatively, such breaks may inactivate the template for amplification. Therefore the quantity of amplified product is dependent on the quality of extracted nucleic acids. Though large molecular weight DNA is frequently obtained from paraffin blocks, the extend of degradation is difficult to quantify. It is therefore important to include internal controls such as amplification of an endogenous gene sequence of about the same size and to verify the PCR product by Southern blot hybridization. The advantages and limitations of this technique are further discussed in Chapter 3.5.

In summary, the isolation of nucleic acids from fixed and paraffin-embedded specimens is a powerful tool, making large numbers of well studied samples accessible for analysis with molecular biology methods, such as filter hybridization and DNA amplification. The method has proved highly useful in retrospective studies of viral pathogenesis. The key limitation is the degree of degradation of nucleic acids which may be partially assessed at the outset by simple histological survey. When conditions and tissues are carefully chosen, reproducibile qualitative results may be obtained.

References

51. Goetz, S. E., Hamilton, S. R., and Vogelstein, B. (1985). *Biochem. Biophys. Res. Commun.* **130**, 118.
52. Shibata, D., Martin, W. J., and Arnheim, N. (1988). *Cancer Res.* **46**, 2964.
53. Löhr, M. and Oldstone, M. B. A. (1989). Submitted.
54. Nerenberg, M. I., Wiley, C. A., Devaney, K., Kramer, B. S., Takebayashi, K., Nagashima, K., Akai, K., and Oldstone, M. B. A. (1989). Submitted.
55. Chirgwin, J. M., Pryzybyla, A. E., MacDonald, R. J., and Rutter, W. J. (1979). *Biochemistry*, **18**, 5294.
56. Rupp, G. M. and Locker, J. (1988). *Biotechniques*, **6**, 56.
57. Maniatis, T., Fritsdi, E. F., and Sambrook, J. (1982). In *Molecular Cloning*, Cold Spring Harbor, New York.
58. Seto, E. and Yue, T. S. B. (1987). *AM. J. Pathol.*, **127**, 409.
59. Schochetmann, G., Du, C. Y., and Jones, W. K. (1988). *J. Infect. Dis.*, **158**, 1154.
60. Osle, C. (1988). *Biotechniques*, **6**, 162.
61. Löhr, M., Nelson, J. A., and Oldstone, M. B. A. (1989). Submitted.
62. Shibata, D. K., Arnham, N., and Martin, W. J. (1988). *J. Exp. Med.*, **167**, 225.

63. Impraim, C. C., Saiki, R. K. Erlich, H. A., and Teplitz, R. L. (1987). *Biochem. Biophys.*, **142**, 710.
64. Lipkin, W. I., Battenberg, E. L. F., Bloom, F. E., and Oldstone, M. B. A. (1988). *Brain Res.*, **451**, 333.

2D. DNA AMPLIFICATION/POLYMERASE CHAIN REACTION
MATTHIAS LÖHR and MARIA S. SALVATO

21. Introduction

DNA amplification via the polymerase chain reaction (PCR) is receiving increased attention because of its high degree of sensitivity, although problems with contamination can cause concern.

Since the technique is covered in great depth elsewhere (66–70), we will focus on our use of the technique on nucleic acids obtained from paraffin-embedded tissues.

22. PCR selection and preparation of primers

Primers (oligonucleotides of 20–30 bases) should be selected by the following criteria: first, it is preferable to have a high G/C content for a higher melting temperature (71), and second, oligonucleotides should not have internal complementarity which would promote self-annealing. In general, it is a good idea to choose primers from conserved areas of a sequence so that they can be used on several genetically related templates.

The ideal primers will result in a 150–300-bp product (including primers). If the template sequence permits, restriction enzyme sites should be included in the 3' and 5' ends of the PCR product to allow visualization of precise products, easier cloning, and in some applications, removal of primer sequences. Alternatively, non-annealing overhangs can be created consisting of restriction enzyme sites. When designing primers, it should be kept in mind that the (*taq*) polymerase extends only in the 5' direction (3'→5').

Oligonucleotide primers should be resuspended in sterile water or TE buffer (72) at 1 mg/ml (μg/μl); for a 20 mer 1 OD_{260} unit = 30 μg. 1–5 OD units can be used in the cartridge purification.

The calculations for the molecular weight of oligonucleotides are as follows:

$$OD_{260} \times \text{dilution} \times 20 \text{ (ext. coeff.)} = \mu g/ml$$
$$(\text{length of oligo}) \times (330 \text{ g/mol of base}) = g/mol$$

For example, for a 20 mer, it is 6 600 g/mol or 6.6 μg/nmol. Thus, there are 140 pmol per μg.

For the PCR, between 1 μg–500 ng or 500 pmol of each primer will be used per reaction.

23. Preparation of DNA

The preparation of DNA may follow standard extraction techniques employing proteinase K treatment and phenol/chloroform extractions (73, 74) or as outlined here. After resuspending the DNA in TE, treat a 100-μl volume of DNA with 5 μl of 10 mg/ml DNase-free RNase for 45 min at 37 °C, followed by another phenol/chloroform extraction, ethanol precipitation (with 0.1 M NH_4Ac) and resuspension in sterile water. After OD_{260} measurement, adjust an aliquot of the DNA to a concentration of 100 ng/μl.

24. DNA amplification

The assembly of a PCR sample should take place in a clean place in the laboratory, physically apart from the space where the PCR products are handled. Keep components on ice and use fresh sterile pipette tips for each addition. The *taq* polymerase should be taken out of the freezer (-20 °C) just prior to use, spun briefly and pipetted slowly.

The sample is to be overlaid with 100 μl of sterile mineral oil (slow pipetting) and vials are closed tightly before transfer to the PCR machine. Before inserting into the temperature blocks, mineral oil should be added to the wells to facilitate better temperature conductivity. For a given set of template DNA and primers, it might be worthwhile to optimize the conditions. With a positive control as template DNA, start out with an excess of the following components: primers from 1 μg, *taq* polymerase from 5 U/assay, and DNA from 0.5 μg. Then titrate (Mg^{2+}) between 1–3 mM. When the optimum concentration is determined, titrate out *taq* polymerase. Next, increase the annealing temperature to 80% of T_{max}. Run 20–30 cycles for each of these experiments.

A typical run would be 20–30 cycles consisting of three steps:
(a) denaturing   94 °C 1 min–1 min 30 sec
(b) annealing   37–55 °C 1–2 min
(c) extension   72 °C 2–3 min

The annealing temperature depends on the T_{max} (i.e. G/C content) of the oligonucleotides. To start out, a lower temperature should be chosen. The time periods for the three steps have to be adjusted according to the delay between the three temperatures (usually $\simeq$ 1–2 min) and the temperature at which the PCR machine is counting the time, e.g. starting to count for 72 °C when sample temperature is 71 °C. In addition, the programme can be extended in that the initial denaturation should total 2 min and the final extension step (after 20–40 cycles) can be a total of 10 min. After the reaction, samples should be kept at 4 °C until further processing. Some programmes offer auto-extension to allow extra time for extension with an increasing amount of DNA (increasing cycles).

The analysis of the PCR products should consist of an agarose or

DNA amplification via PCR

polyacrylamide gel on which 1/10–1/5 (10–20 μl) of the PCR products should be loaded. In addition, there is another advantage in that the gel can be directly used for Southern transfer using a rapid transfer system to Zeta bind (75) or vacuum blotting and probed with a labelled oligonucleotide from the middle portion of the amplified segment. Alternatively, slot blot or dot blot analysis can be employed, which is particularly suitable for large numbers of samples.

24.1 PCR general considerations

For a given DNA amplification system, it is worthwhile to optimize the conditions. The protocol is as follows:

Protocol 16. Optimal conditions for DNA amplification

Excess of
- primers (1 μg)
- *taq* (5 units/assay)
- DNA (0.5 μg), annealing at 37 °C

1. Titrate (Mg^{2+}) between 1–3 mM.
2. When optimum concentration is determined, titrate out *taq* polymerase.
3. Increase annealing temperature to 80% of T_{max}. Run 20–30 cycles for each of these experiments (see below).

24.2 PCR

- Keep all solutions, pipette tips, etc. separate from all other solutions.
- Have special siliconized tubes (Robbins Scientific), sterile H_2O and pipette tips solely for use in PCR.
- Assemble the reaction mix (see *Protocol 17*), in the tissue culture hood or a clean benchtop.
- Typical volume of a PCR is 50–100 μl in a small Eppendorf tube.
- Change pipette tips every time to avoid cross contamination.
- Keep everything on ice.
- *Taq* polymerase should be taken out of the freezer (−20 °C) just prior to use; spin briefly (touch); pipette slowly.

Overlay the sample with 100λ of mineral oil (slow pipetting) and close vials tightly. Transfer to PCR machine. Before inserting into holes, add some mineral oil into the wells ($\simeq 100\lambda$). Before starting the machine, check the programme you want to run.

A typical PCR run would consist of:
(a) denaturing 1 min 30 sec at 94 °C
(b) annealing   2 min at 37 °C (or 55 °C)
(c) extension   3 min at 72 °C

Protocol 17. Set-up for PCR

Reagents, to be added in the following order:	Quantity (μl)
1. Double distilled, sterile H_2O	56
2. Reaction buffer (5 ×)	20
3. 0.5 M βME (or 0.1 M DTT)	2(10)
4. dNTPs mix	16
5. Primer 1	2
6. Primer 2	2
7. template DNA	1
8. *taq* polymerase	1
	100

25. Cloning and sequencing of PCR products (76, 77)

To check the sequence of a PCR product, it can be cloned into a vector and checked by direct double-stranded sequencing.

If no suitable restriction enzyme sites are available, ligation into a blunt end restriction site of the vector can be employed, i.e. *Hinc*II in pUC19. For direct sequencing of the PCR reaction products, run the PCR reaction as usual. Isolate the amplified band from an agarose gel. Use 1–2 μg of the PCR product. Either one of the primers initiating the PCR reaction can be used as a primer but primers from within the amplified sequence would be preferable. Follow the procedures of one of the commercially available sequencing kits, such as the Sequenase kit from Pharmacia.

26. RNA as a template for PCR

For certain questions, such as searching for an RNA virus or transcription of genes (mRNA) (88), it is necessary to detect low or single copies of RNA rather than DNA. Since the *taq* polymerase only works with DNA, this can only be achieved by using a reverse transcriptase producing a cDNA, which will serve as a template for the subsequent PCR. *Protocol 18* was originally worked out and successfully used with lymphocytic choriomeningitis virus RNA (76).

Protocol 18. RNA for PCR

Use between   1–4 μg (purified viral) or
              4–8 μg (total cellular) RNA.
  1. Add primers, 500 pmol

Protocol 18. *Continued*

2. Boil for 1 min 30 sec (NO CHILL)
3. Incubate for 45 min at 42–45 °C (annealing)
4. Add the following; in order
    - (a) X µl   Sterile H_2O
    - (b) 20 µl   5 × RT^a buffer
    - (c) 2 µl   RNAsin
    - (d) 10 µl   0.1 M DTT
    - (e) 6 µl   20 mM dNTP's
    - (f) 4 µl   Reverse transcriptase
5. Adjust to 100 µl
6. Incubate for 15 min at 42–45 °C
7. Stop reaction with 4.5 µl of 10 mM EDTA (make fresh: 20 µl of 0.5 M EDTA + 980 µl of H_2O)
8. Boil for 1 min 30 sec
9. Incubate for 45 min at 42 °C
10. Touch spin
11. Add *taq* polymerase (1 µl = 5 units)
12. Add mineral oil (100 µl)
13. Go to PCR, skip the first 94 °C cycle (start with annealing)
14. Run 30–40 cycles

[a]5 × RT: 250 mM Tris–HCl pH 8.3
             250 mM KCl
             30 mM $MgSO_4$

References

66. Saiki, P. K., Scharf, S., Faloona, F., Mullis, K. B., Horn, G. T., Erlich, H. A., and Arnheim, N. (1985). *Science*, **230**, 1350.
67. Mullis, K. B. and Faloona, F. A. (1987). *Methods Enzymol.*, **155**, 335.
68. Crescenzi, M., Set, M., Herzig, G. P., Weiss, P. D., Griffiths, R. C., and Korsmeyer, S. J. (1988). *Proc. Natl. Acad. Sci. USA*, **85**, 4869.
69. Oste, C., (1988). *Biotechniques*, **6**, 162.
70. Gyllenstein, U. B. and Erlich, H. A. (1988). *Proc. Natl. Acad. Sci. USA*, **85**, 7652.
71. Lathe, R. (1985). *J. Mol. Biol.* **183**, 1.
72. Mihovilovic, M. and Lee, J. E. (1989). *Biotechniques*, **7**, 14.
73. Newton, C. R., Kalsheker, N., Graham, A., Powell, S., Gammack, A., Riley, J., and Markham, A. F. (1988). *Nucleic Acids Res.*, **16**, 8233.
74. Lo, Y. M. D., Mehal, W. Z., and Fleming, K. A. (1988). *Nucleic Acids. Res.*, **16**, 8719.
75. Joly, E., Salvato, M., Whitton, J. L., and Oldstone, M. B. A. (1989). *J. Virol.*, **163**, 1845.

76. Salvato, M. S. and Shimomaye, E. M. (1989). *Virology*, **173**, 1.
77. Löhr, M., Nelson, J. A., and Oldstone, M. B. A. (1989). *AAP Transactions*, **CII**, 213.
78. Gait, M. J. (ed.) (1984). *Oligonucleotide Synthesis: A Practical Approach*. IRL Press, Oxford.
79. Maniatis, T., Fritsch, E. F., and Sambrook, J. (1982). *Molecular Cloning. A Laboratory Manual*. Cold Spring Harbor Laboratory.
80. Glover, D. M. (ed.) (1987). *DNA Cloning Volume III: A Practical Approach*, IRL Press, Oxford.
81. Goetz, S. E., Hamilton, S. R., and Vogelstein, B. (1985). *Biochem. Biophys. Res. Commur.*, **130**, 118.
82. Dubeau, L., Chandler, L. A., Gralow, J. R., Nichols, P. W., and Jones, P. A. (1986). *Cancer Res.*, **46**, 2964.
83. Imprain, C. C., Saiki, R. K., Erlich, H. A., and Teplitz, R. L. (1987). *Biochem. Biophys. Res. Commur.*, **142**, 710.
84. Shibata, D., Martin, W. J., and Arnheim, N. (1988). *Cancer Res.*, **48**, 4564.
85. Toneguzzo, F., Glynn, S., Levi, E., Mjolsness, S., and Hayday, A. (1988). *Biotechniques*, **6**, 460.
86. Todd, J. A., Bell, J. E., and McDevitt, H. O. (1988). *Nature*, **329**, 599.
87. Loh, E. Y., Elliot, J. F., Cwirla, S., Lanier, L. L., and Davis, M. M. (1989). *Science*, **243**, 217.
88. Biggin, M. D., Gibson, T. J., and Hong, G. F. (1983). *Proc. Natl. Acad. Sci. USA*, **80**, 3963.

2E. TECHNIQUES FOR DOUBLE-LABELLING VIRUS INFECTED CELLS
CATHERINE REYNOLDS-KOHLER and JAY A. NELSON

27. Introduction

Our understanding of viral pathogenesis has increased dramatically over the past few years with the development of techniques and specific probes to detect viruses directly in tissue. Analysis of DNA and RNA extracted from biopsy and autopsy specimens by Southern and Northern blotting procedures are useful methods for determining the presence of virus. These techniques in combination with specific sequence amplification by the polymerase chain reaction (PCR) (see Chapter 2D) have proven to be an exquisitely sensitive way to detect low copy numbers of viruses in these materials. Although PCR technology has temendous potential, identification of infected cell types remains a limitation with this procedure. For this aspect of viral pathogenesis *in situ* hybridization and immunocytochemistry are useful techniques to localize viruses and viral gene products in individual cells. Utilizing combinations of these procedures cell types can unambiguously be identified in these tissues. The purpose of this chapter is to describe technical aspects of double-labelling techniques. We will first discuss methods utilized for immunocytochemistry and *in situ* hybridization separately followed by double-labelling procedures. Although alternative methods exist for these procedures, techniques described in this section are relatively straightforward and are currently used in our laboratory (89–93).

28. Preparation of slides and specimens

The slide and specimen preparation discussed in this section applies to both immunocytochemistry techniques and *in situ* hybridization. The slides must first be acid cleaned, followed by a coating of poly-D-lysine. The polylysine coating allows the section or cells to attach readily to the surface of the slide. Failure to precoat the slides can result in the loss of specimens from the slide during the following procedures (see *Protocol 19*).

Protocol 19. Coating slides with poly-D-lysine

1. To clean glass slides, soak them in chromic acid for 4–24 h. Wash the slides briefly with tap water and then continue to soak in tap water for 4 h. Rinse the slides briefly in deionized water three times, 2 min each.
2. To coat the slides, make a 0.1% poly-D-lysine (Sigma) solution with deionized distilled water (ddH_2O). Store this solution at $-20\ °C$ and thaw at $37\ °C$ before each use. The solution can be used to coat approximately 1000 slides before a new polylysine solution must be made up.
3. Allow the acid-cleaned slides to soak in the polylysine solution for 30 min, then rinse the slides quickly three times in ddH_2O.
4. Dry the slides by placing them in a warm room at $37\ °C$.
5. After the slides are dry, bake them in a vacuum oven at $80\ °C$ for 1 h. It is important not to omit the last step, since the baking ensures properly coated slides. The slides can then be stored indefinitely at room temperature in a clean covered slide rack.

Prior to attaching tissue or cells to polylysine coated slides, the specimens should be fixed or frozen. Any animal or human tissue obtained during a biopsy or autopsy should be fixed or frozen immediately to minimize the degradation of nucleic acids and proteins. Tissue culture cells or fresh cell isolates should be applied to a polylysine coated slide and fixed after the cells are completely dried to the surface.

A variety of different fixatives are available. Both the fixative and the fixation process vary with the permeability of the tissue or cells. Fixatives such as ethanol–acetic acid and methanol–acetone allow efficient hybridization during the *in situ* procedure, yet there is a lack in the preservation of morphology. The preservation of tissue and cell ultrastructure is an important aspect of *in situ* hybridization, but retention of morphological detail is imperative for immunocytochemistry. An alternative group of fixatives are the crosslinking agents. This list includes formaldehyde, glutaraldehyde and paraformaldehyde-lysine-periodate (PLP) (94). These fixatives are superior in preserving morphology and seem not to interfere with hybridization efficiency when used at concentrations of 1–4% of the total volume (95).

After tissues are fixed they are embedded into paraffin blocks. The blocks are cut into 8–10 µm sections. Float the sliced sections in a 45 °C distilled water bath containing 1% Elmers white glue. With the polylysine coated slides, pick up the sections and arrange them on the slide. If the specimen is small, multiple sections can be mounted on the slide. If the section is larger than 15 mm only one section should be mounted per slide. Air-dry the sections overnight and keep stored in a dry slide holder at 4 °C. Tissue embedded in paraffin this way will be good indefinitely. Failure to use the correct amount of Elmers glue, or the use of slides not coated with polylysine, will result in partial or total loss of tissues during the procedures (96).

For frozen tissue, the procedure is as follows: fix the tissue by perfusing it with 2% paraformaldehyde at room temperature then soak it in a 15% sucrose/PBS solution at 4 °C for 1 h for further cryoprotection.

Tissue tek medium is used as the freezing medium for the fixed sections (see *Protocol 20*).

Protocol 20. Embedding fixed tissue in tissue tek

1. Place an amount of tissue tek medium on to a gold cryostat stub that is roughly equivalent to the size of a nickel. Allow the tissue tek to become partially firm.
2. Orient the fixed tissue in the medium and apply more medium to cover the sides and the top of the specimen. The specimen is frozen and flattened by the application of the cold sink on top of the tissue.
3. After the tissue is sufficiently frozen, trim away excess tissue tek and form a rectangle.
4. From these frozen sections cut 7-µm thick slices and mount as above on to polylysine coated slides.

Cells from peripheral blood are optimally preserved when fixed with a 2% PLP. The preparation of 2% PLP is shown in *Protocol 21*.

Protocol 21. Preparation of periodate-lysine-2% paraformaldehyde fixative

Materials
- lysine HCl

Protocol 21. *Continued*
- Na_2HPO_4
- $NaH_2PO_4 \cdot H_2O$
- paraformaldehyde
- Na periodate

Method
1. Prepare Na phosphate buffer [a] pH 7.4: dissolve 22.8 g of Na_2HPO_4 and 5.46 g of NaH_2PO4 in 2 litres of millique water.
2. Dissolve 36.6 g of lysine HCl[a] in 1 litre of millique water (0.2 M solution), and titrate to pH 7.4.
3. Prepare 8% paraformaldehyde[a]: place 8 g of paraformaldehyde into 10 ml of millique water over a heat/stir apparatus. Bring the solution to clarity with NaOH then add an additional 90 ml of millique water to attain 8% concentration.
4. Add 1 litre of Na phosphate buffer to the lysine HCl solution (0.1 M solution).
5. Add 100 ml of 8% paraformaldehyde to 300 ml of 0.1 M lysine HCl/Na phosphate buffer solution.
6. Add 0.9 of Na periodate to the paraformaldehyde/lysine HCl/Na phosphate buffer solution (0.01 M solution).
7. Bring to pH 7.4.
8. Microfilter just prior to use to ensure sterility.

[a] The phosphate buffer, the lysine HCl buffer and the paraformaldehyde solution can all be made beforehand and stored. The Na periodate is the critical reagent to be added fresh each time the solution is made.

Protocol 22. Attachment of cells to polylysine coated slides
1. First isolate the cells by Ficoll–Hypaque gradients and adjust to a concentration of 1×10^5–1×10^7 cells/ml, depending on the cell size.
2. Arrange the cells on polylysine coated slides in 5–10 μl dots.
3. After the cells are dotted, dry the slides for 1 h. Continue the procedure by fixing in 2% PLP for 15 min (*Protocol 21*).
4. Gently rinse the slides twice in PBS for 2 min each.
5. Dehydrate the slides in graded washes of ethanol: 30%, 50%, 70%, 95% and 100%, for 1 min each.
6. Store the cells in a clean, dry, covered slide rack at 4 °C. If fresh PLP is made up before the fixation, the slides should be good for 3 weeks.

29. Immunocytochemistry

After the sections or cells have been properly fixed and attached to polylysine coated slides, immunocytochemical techniques may be applied. Cell morphology is best preserved with fixed tissue embedded in paraffin. However, in our experience, few antibodies work well in paraffin-embedded tissue.

Immunoreactive sites on tissues and cells are detected by using specific antibodies directed against viral polypeptides or cellular proteins. These antibodies can be polyclonal or monoclonal. The specificity of the antibody is important in reducing the chance of high background, due to non-specific binding.

Most antibodies are commercially available and can be obtained from: Dako Immunoglobulins (Copenhagen, DK); Vector Laboratories Inc. (Burlingame, CA); or Tago (Burlingame, CA). Instructions for use and recommended dilutions are usually included in the package.

29.1 Detection systems

29.1.1 Immunoperoxidase techniques

The two most widely utilized immunoperoxidase detection systems are the Avidin–Biotin Complex (ABC) and the Peroxidase–anti–Peroxidase (PAP) complex. The end result in both of these techniques gives rise to an antibody linked to peroxidase. The peroxidase, in the presence of hydrogen peroxide, reacts with a chromagen to produce a stain. The stain represents specific antigenic sites on the tissue or cells. The Avidin–Biotin Complex (ABC Kit from Vector Labs) is based on the following principles: avidin, an egg white glycoprotein, possesses four active sites for binding biotin, a water soluable vitamin. Therefore, avidin with its extraordinarily high affinity for biotin can react with biotinylated molecules such as antibodies and horseradish peroxidase. The end result will be the formation of a highly stable complex that contains multiple peroxidase molecules. The general scheme for this technique is as follows:

(a) The primary antibody, specific for a particular antigenic determinant is applied to the slide (for a two-step method, the primary antibody can be conjugated to biotin).

(b) If the primary antibody is not biotinylated, then a secondary biotinylated antibody against the primary antibody is applied to the slide.

(c) The Avidin–Biotin (Peroxidase) complex is applied to the slide.

(d) A chromagen is reacted with the Avidin–Biotin complex, catalysed by H_2O_2, to form a stain.

The PAP system (Jackson ImmunoResearch Lab Inc.), utilizes a different approach to achieve the same outcome as in the ABC technique. The difference being the absence of biotinylated antibody in this procedure. Instead, the

Techniques for double-labelling

secondary antibody serves as a bridging antibody between the primary antibody and the PAP complex. The source of the bridging antibody should be from antisera to immunoglobulin from the same host species as the PAP complex. The exposed peroxidase molecules can then be reacted with chromagens in the presence of H_2O_2.

Two commonly used chromagens for peroxidases are (i) 3,3' diaminobenzadine (DAB), and (ii) 3 amino-9 ethylcarbazole (AEC). DAB and AEC are both potential carcinogens and should be handled with appropriate perecautions. (The recipe for making the two chromagens is given in Section 29.2).

29.1.2 Non-peroxidase detection techniques

There are two main non-peroxidase detection systems utilized; the avidin–alkaline phosphatase (AAP) and the glucose oxidase–anti-glucose oxidase (GAG) complexes. These procedures technically parallel the two immunoperoxidase methods.

The avidin–alkaline phosphatase forms an immune complex with a biotinylated secondary or primary antibody. The difference in this technique is that the enzyme alkaline phosphatase is reacted with a different chromagen than used with the peroxidase complexes. AAP reacts with Fast Blue which contains napthal phosphoric acid (PO_4) and 0.1 M Levamisole. (The recipe is included in Section 29.2).

GAG complexes are formed in the same fashion as the PAP complexes. A secondary bridging antibody links the primary antibody with the GAG complex. The glucose oxidase molecule is reacted with either DAB or AEC in the presence of H_2O_2 to form a stain. One advantage of the glucose oxidase system is the fact that the enzyme is not present in mammalian tissue. Since it is a fungal enzyme, no pretreatment of tissues or cells to inhibit endogenous enzyme activity is necessary.

When deciding which detection system is best to use, several factors must be taken into consideration. The first important factor is the tissue type. Different tissue types will have varying permeability and varying concentrations of endogenous enzymes that may interfere with the detection process. Another important factor is the amount of antigen present. The extent of amplification needed to detect antigen will depend on its concentration in the cells or tissues being tested.

The following lists some important information that will help in deciding which detection system should be used.
- Certain tissue types, especially liver and kidney, are rich in biotin. Therefore, endogenous biotin must be blocked (97).
- Avidin binds to lectins, particularly in neural tissue and must be blocked. Pre-incubation of tissue sections with α-methyl-D-mannoside may be necessary to reduce non-specific staining (98).
- Intestinal and placental tissue contains high endogenous levels of alkaline

phosphatase. Use one of the peroxidase complexes to aleviate non-specific background.
- The ABC detection system can be used for all systems and is only contingent upon a biotinylated antibody. The PAP method requires different PAP complexes depending on the species source of the primary antibody.
- Non-specific background can also be decreased if the primary antibody is biotinylated. This two-step method is not only faster, but yields a sharper and more well defined staining result. Best results of this two-step method occur when used in conjunction with DAB.
- If the slides will be under constant viewing and/or are going to be stored for a long time, fading might be a problem. The PAP system should not be used due to the fading characteristics of the peroxidase reactions. However, the ABC, specifically used in conjunction with DAB is very stable. Also the chemical activity of glucose oxidase in the GAG system forms very stable compounds.
- If the slides are going to be further processed after staining, it may be important that the stain is not alcohol soluble. DAB is a good stain to use, especially during immunocytochemistry–*in situ* double-labelling technique because the latter technique requires dehydration steps.
- For double-label immunocytochemistry, make sure that the two stains are going to be identifiable after the procedure. Slides stained with both DAB and AEC make it hard to distinguish specific morphological areas.

29.2 Staining

The following consists of information regarding the preparation and development of the stains employed in the immunocytochemical techniques previously described.

Protocol 23. DAB

Preparation

This reagent comes as a powder and is reconstituted at a concentration of 1 mg/ml with PBS. The solution can be stored in 5–10 ml aliquots at $-20\,°C$ for a couple of months. A purple–brown precipitate may form as the solution is made. This is normal and the amount of precipitate will increase as the frozen stock gets older.

Development

Thaw the DAB and dilute to 0.5 mg/ml with PBS prior to use. Add 1.7 μl of 30% H_2O_2 for every 5 ml of DAB and draw the DAB/H_2O_2 mixture into a syringe. Before using DAB on a slide, filter the stain through a 0.45-μm syringe filter to get rid of most of the precipitate. Watch the development under a microscope. The reaction will take any time from 3 to 10 min. If no colour has developed after 10 min, the reaction can be pushed at 37 °C for 5 min. (If no colour has appeared

Protocol 23. Continued

by this time, the reaction did not work). Stop the reaction by immersing the slides in PBS, then wash three times, for 5 min each, with PBS.

DAB stains the sections with a reddish–brown colour. Since the stain is not alcohol soluble, the slides should be dehydrated in gradual ethanol washes and then mounted with a coverslip.

Protocol 24. AEC

Preparation

The AEC solution used during staining should be made up fresh for each experiment. However, a stock solution of AEC concentrate can be stored at 4 °C for 2–3 months (AEC is light-sensitive so be sure to store the solution in a dark bottle).

(a) Stock solution: add 0.1 g of AEC to 25 ml of N-N' dimethylformamide.

(b) Working solution
 - 1 ml of AEC stock solution
 - 14 ml of 0.1 M acetate buffer, pH 5.2
 - 15 μl of 30% H_2O_2 (hydrogen peroxide)

[to make 0.1 M acetate buffer pH 5.2: add 210 ml of 0.1 N acetic acid (5.75 ml of glacial acetic acid in 1 litre of H_2O), to 790 ml of 0.1 M sodium acetate (13.61 g of sodium acetate trihydrate in 1 litre of H_2O)]

Development

Equilibrate the slides in 0.1 M acetate buffer for 5 min prior to staining. During the 5 min, make up the working solution (if the working solution sits for longer than 30 min before use, the AEC will start to precipitate).

The H_2O_2 should be added right before the staining procedure. After the addition of H_2O_2, the stain should be filtered through a 0.45-μm syringe filter unit to remove any residual precipitate. The solution should cover the entire section and remain on the slide for 20 min. Due to non-specific staining, the entire section may turn a pinkish–red colour during the 20-min incubation time. This is normal, and the slides need only be monitored under a microscope after the 20 min time period has expired. If the sections still lack specific staining, place them at 37 °C for a maximum of 10 min. Monitor the staining under the microscope in 3-min intervals since the stain intensifies significantly at 37 °C. To stop the reaction rinse the slides in PBS, then continue to wash them three more times in PBS, for 5 min each.

AEC stains the sections with a reddish–pink stain. This stain is *NOT* alcohol soluble. Let the slides air-dry before mounting a coverslip.

Protocol 25. Fast Blue

Preparation

This stain should be made up fresh prior to use. Solutions I and II are made up as follows:

Solution I: 10 mg of napthal phosphoric acid in 4 drops of dimethylformamide.

Solution II: 10 mg of Fast Blue in 20 ml of 1 M Tris buffer (pH 9.2).

Development

Equilibrate the slides in 1 M Tris (pH 9.2), for 5 min prior to staining. In the meantime mix solution I and solution II together and then add 400 µl of 0.1 M Levamisole (the Levamisole is diluted with ddH_2O). Once the two solutions are combined and the Levamisole is added, the mixture will start turning from a yellow colour to a brown colour. This reaction can happen in a matter of minutes, so the stain must be used immediately. This mixture should be filtered with a 0.45-µm syringe filter before applying to the slides. The Fast Blue stain should appear within 3 min. Stop the reaction by immersing the slides in PBS, then continue to rinse three times for 5 min each.

Fast Blue, as the name suggests, gives rise to a stain that is bluish–purple in colour. Fast Blue is not alcohol soluble. Therefore, the slides should be dehydrated in gradual ethanol washes before being coverslipped.

29.3 Immunocytochemistry: a method for single-labelling

This technique—described in *Protocol 26*—utilizes a single antibody to identify a specific antigenic site.

Protocol 26. Immunocytochemistry: single-labelling method

1. If the tissue or cells have been embedded in paraffin before adherence to polylysine coated slides, the paraffin must first be dissolved with xylene or histoclear. The tissue should be washed twice, 5 min each with either solvent.

2. Clear the slides of xylene or histoclear by washing in ethanol for 2 min.

3. To block endogenous peroxidases wash the slides for 30 min with 0.3% H_2O_2 in methanol. (This step is only necessary when using a detection system that utilizes peroxidase).

4. Gradually hydrate the slides in sequential washes of 100%, 95% and 70% ethanol. Once they have been hydrated, it is important to keep the slides from drying out.

5. Wash the slides three times, 5 min each in PBS.

Protocol 26. *Continued*

6. To maximize antibody–antigen complex formation, accessibility to the antigenic site must be achieved. A proteinase K treatment for 15 min at 37 °C allows sufficient digestion of the tissue without morphological destruction. Only tissue culture cells should be treated with proteinase K. Fresh cell isolates are too fragile and the proteinase K digestion would destroy the cells. The following lists the proteinase K concentration and buffers used for tissue sections and cells in culture.

Tissue	Cells
10 μg/ml proteinase K	1 μg/ml proteinase K
50 mM Tris pH 7.0	50 mM Tris pH 7.0
2 mM $CaCl_2$	2 mM CaCl

    The concentration of proteinase K needed to achieve sufficient digestion with a minimum amount of morphological destruction will vary from tissue to tissue. In some cases antigens may be sensitive to protease digestion. In general, most tissue will need between 1–20 μg of proteinase K, yet there may be some that requires as much as 50 μg. When first attempting to stain a particular tissue type, try varying concentrations of proteinase K to determine the ideal concentration needed.

7. After digestion, non-specific antibody reactions are prevented by using a 10% solution of blocking serum in PBS. It is important that the serum used for blocking is from a species of animal that is unrelated to the animal the primary antibody was raised against. Normal swine serum is a good blocking agent. The blocking serum should cover the section entirely and stay on the slide for 30 min at room temperature.

8. After 30 min, shake off the blocking serum into a Kimwipe (do not rinse). Add the primary antibody diluted approximately 1:50 in PBS with 1% blocking serum. Add enough of the primary antibody so that the section will be entirely covered when coverslipped. The amount of primary antibody necessary will depend on the size of the section. Make sure there is a sufficient amount of antibody solution so that the slides don't dry out.
    - If the primary antibody is a monoclonal, then the reaction takes place at either 37 °C for 2 h or 4 °C overnight.
    - If the primary antibody is a polyclonal, the reaction takes place at 37 °C for 2 h.

9. Either the next day (monoclonal primary antibody) or 2 h later (polyclonal primary antibody) remove the coverslips by immersing the slides in PBS. The coverslips should fall off the slide. Be careful not to pry the coverslips off because the tissues may come off the slide. Continue to wash three times, 5 min each in PBS.

10. The secondary antibody used in the procedure will depend on which detection system will be applied. The secondary or primary antibody (two-

Protocol 26. *Continued*

step procedure), must be biotinylated if any of the avidin–biotin detection systems are to be used. If either the PAP or GAG systems are to be utilized, the secondary antibody serves as the bridging antibody. The secondary antibody should be from antisera to IgG from the same host species as the PAP or GAG complexes. In general, the secondary antibody should be diluted 1/50 and the antibody should entirely cover the section. The antibody remains on the slides for 30 min at room temerpature.

11. After the incubation period is over, rinse the slides three times, 5 min each in PBS.
12. Once again, this step varies depending upon which detection system is to be utilized.
    (a) For a biotinylated secondary antibody, the Avidin–Biotin Complex (Vector Labs), or the Avidin–Alkaline Phosphatase should be applied to the slides. The AAP is diluted 1/10 with PBS and the ABC is prepared as follows:
        i. In an Eppendorf tube combine 12 μl of avidin and 12 μl of biotin.
        ii. Add 1 ml of 2 × SSC (see *in situ* protocol for recipe).
        iii. This solution must incubate on ice for 30 min before it is added to the slide.
    (b) Apply the AAP or ABC to the slide for 30 min at room temperature.
    (c) PAP or GAG are diluted 1/10 with PBS and added to the slides so that the sections are completely covered. This reaction takes place at room temperature for 30 min.
13. After the complexes have formed and are attached to the sections, any of the previously described staining procedures can be applied. DAB and AEC are to be used with peroxidase complexes and Fast Blue reacts with alkaline phosphatase complexes.

29.4 Immunocytochemistry: a method for double-labelling

This technique allows dual staining by using two different detection systems. The technique is performed exactly as described above for the single label except that there are two primary antibodies and two secondary antibodies. One of the secondary antibodies should be biotinylated and the other should be unconjugated.

After the addition of the two secondary antibodies, perform only one of the detection systems. Once the first stain is deposited on to the slide, proceed with the protocol for the second detection system. When both stains are deposited, let the slides air-dry (or wash with ethanol, if both stains are not ethanol soluble). After the slides are completely dry, coverslip and view.

30. In situ hybridization method

This technique is utilized to screen tissues or cells for the presence of viral nucleic acids. DNA- or RNA-specific fragments are labelled with ^{35}S to be used as radioactive probes to detect these nucleic acids. Other isotopes such as ^{32}P and ^{3}H can be utilized to label fragments, but ^{35}S-labelled nucleic acids are sensitive and give the best signal to noise ratio. The following method describes the technique for conducting *in situ* hybridization with ^{35}S-labelled DNA probes.

DNA probes are prepared by incorporation of deoxyadenine 5'-([α-^{35}S]thio) triphosphate and deoxycytidine 5'-([α-^{35}S]thio) triphosphate (1000 Ci/mM) (Amersham) by nick translation to yield probes which range in size from 50 bp to 200 bp. This size range of ^{35}S-labelled fragments can be optimized by adjusting the concentration of DNaseI concentration. Probes generated in this method should have a specific activity of $> 1 \times 10^8$ c.p.m./μg of DNA. DNA can also be labelled to higher specific activity ($> 1 \times 10^9$ c.p.m./μg) by random hexanucleotide priming (99). However, greater background noise can occur with DNA probes utilizing the latter labelling method. Low melt agarose or polyacrylamide gel purification of probe DNA from the cloning vector prior to labelling is sometimes necessary to decrease non-specific background in specimens such as bowel tissue (93).

In terms of preservation of cell morphology, *in situ* hybridization results are best when performed on paraffin-embedded tissue. Formalin-fixed tissues utilized in this technique seem to be sturdier and hold up better through the rigorous washing steps in the procedure. Cells will readily attach to polylysine coated slides if fresh PLP fixative is used. Remember, cells are only preserved for 3 weeks after they have been fixed on to the slides.

The entire *in situ* procedure takes 6 days. On the first day the slides are pretreated to allow accessibility of the radioactive probe into the tissues or cells. Then, the slides are hybridized with the probe overnight. The next day involves a series of stringent washes to remove bound non-specific probe. The same day the slides are dipped in a photographic emulsion that will develop over a 4-day period. On the sixth day of the procedure, the slides are developed using standard developer and fixer. The 4-day developing period is the most ideal. If the slides are developed after 3 days there is a significant decrease in signal. Conversely, letting the slides go 5 or 6 days prior to developing, will result in high non-specific background.

The probe itself should not be used for more than 2 weeks or else high non-specific background results. This same phenomenon can occur if cells fixed to slides are used after the 3-week time point. Paraffin-embedded tissue can be used indefinitely without any effect on the sensitivity of the probe.

The following protocols are written by days. Before the written protocol, a list of reagents needed for that day is supplied. Recipes for some of the solutions follow the daily instructions.

Protocol 27. *In situ* hybridization method: day 1

Reagents needed
Pretreatment
- xylene
- ethanol washes, 100%, 95%, 70%, 50% and 30%. All washes contain 0.33 M NH_4Ac (11.6 g in 500 ml of $EtOH/H_2O$)
- 0.2 M HCl
- 10% Triton X-100
- 10 mg/ml proteinase K (Boeringer Mannheim)
- 2 M Tris–HCl pH 7.4
- 0.2 M $CaCl_2$
- PBS glycine (2 g/litre)
- PBS

Hybridization
- 100% Formamide (deionized)
- 20 × hybridization salts (see recipe at the foot of this protocol)
- 50 × Denhardt's (see recipe at the foot of this protocol)
- 50% Dextran sulphate—make fresh (0.5 g/ml H_2O—heat at 60 °C)
- 10 mg/ml salmon sperm DNA
- 5 mg/ml HeLa cell extract RNA
- 2000 U/ml clinical heparin
- 10% SDS
- ^{35}S probes (dCTP, dATP)
- 1 M DTT (dithiothreitol)

Procedure
Pretreatment
1. Deparaffinize the tissue sections in 2 × 5 min xylene or histoclear washes.
2. Rehydrate the slides in graded washes of EtOH for 1 min each: 100%, 95%, 70%, 50%, 30%.
3. Rinse twice for 1 min each in distilled deionized H_2O (ddH_2O).
4. Soak the slides in 0.2 M HCl for 10 min. This reagent can be re-used, so save it after the soaking.
5. Rinse the slides twice for 1 min each in ddH_2O.
6. The next step involves treating the slides with Triton X-100. The timing of this step is crucial. Specimens left soaking in Triton X for more than the prescribed time could become entirely digested: dilute 10% Triton X-100 to 1% and wash for 1.5 min at room temperature.
7. Rinse the slides immediately in ddH_2O and then further wash the slides twice for 1 min in PBS.

Protocol 27. *Continued*

8. (This step should be omitted for experiments utilizing fresh cell isolates or tissue cultured lymphocytes. These cells are too fragile and their morphology would be destroyed during the digestion). Similar to the proteinase K digestion during immunocytochemistry, different tissue types will need varying concentrations of proteinase K. If a new tissue type is being tested, the first experiment should include varying concentrations of proteinase K to see which is most optimal. In general, for tissue sections and fibroblasts:
   - 1 μg/ml proteinase K, in
   - 20 mM Tris–HCl, and
   - 2 mM $CaCl_2$
   - for 20 min at 37 °C.

   A few tissue types that work well at concentrations of 10 μg/ml of proteinase K are:
   - Rectal/intestinal tissue
   - Placenta
   - Eye
   - Liver
   - Thymus
9. A 5-min wash in PBS glycine (2 g/litre).
10. Rinse the slides in PBS for 3 min.
11. Dehydrate through ethanol washes: 30%, 50%, 70%, 95%, 100% and then air-dry. (Better morphological preservation is found through the gradual rehydration or dehydration steps).

Hybridization

1. Initially, make up the Dextran sulphate solution since it takes approximately 30 min to go into solution at 60 °C. Make sure to vortex the solution thoroughly before heating.
2. Calculate the total amount of hybridization mix needed. For cells plated down as described in the fixation section, 5 μl of hybridization probe mix per dot of cells will be sufficient. For tissue sections, depending on the size, 10–50 μl of hybridization mix with probe per section will be needed. After calculating the total mix for all the slides in the experiment the mix can be made. See *Table 4* for the hybridization mix.
3. The final probe concentration in the hybridization mix should be approximately 10^5 c.p.m./μg. The probe is made beforehand at a concentration of approximately 2×10^6 c.p.m./μg. At 2×10^6 c.p.m./μg, the probe will be 5% of the total hybridization mix to reach a final concentration of 10^5 c.p.m./μg. The probe is stored at -70 °C and should not be used for more than 2 weeks.
4. The probe hybridization mix is denatured by boiling for 2 min and then quenched on ice to keep from re-annealing.

Protocol 27. *Continued*

Table 4. Hybridization mix

Reagent	Stock	Final concentration	Amount for 200 μl
Formamide (deionized)	100%	50%	100 μl
Hybridization salts	20×	5×	50 μl
Denhardt's[a, b]	50×	5×	20 μl
Dextran sulphate—make fresh (0.5 g/1 ml H$_2$O—heat at 60 °C)	50%	10%	20 μl
Salmon sperm DNA[b]	10 mg/ml	500 μg/ml	5 μl
HeLa cell RNA[b]	5 mg/ml	250 μg/ml	5 μl
Heparin—clinical[b]	2000 U/ml	20 U/ml	2 μl
SDS	10%	0.1%	2 μl

[a] See Maniatis manual.
[b] Stock solutions can be made up ahead of time, aliquoted and stored at −20 °C. The rest of the stock solutions, except dextran sulphate which must be made up fresh, can be made up and stored at room temperature.

5. Add DTT to a final concentration of 10 mM to each probe. DTT is a reducing substance and keeps the ^{35}S from forming disulphide bonds. Failure to add DTT to the probe mix will result in failure to see any signal at all. This step is very crucial.

6. Add the probe mix to the sections or cells. The best result for applying probe mix to tissues is in 5-μl aliquots. This allows every part of the specimen to be sufficiently covered with probe when a coverslip is applied to the slide. Be sure to have enough probe mix on your specimen so that no part is left dry. If too little mix is applied, the tissue will tend to lift off the slide the next day when the coverslip is removed. Conversely, don't put too much mix on the slide so that the probe mix will run out under the coverslip.

7. Coverslip specimens with Gel-Bond film that has been cut to fit the section. Gel-Bond film is best to use because of its flexibility and its hydrophobic/hydrophilic quality. Check the Gel-Bond with a drop of H$_2$O to see which side is hydrophilic and which is hydrophobic. Mount the coverslips so that the hydrophobic side faces the specimen.

8. Seal the coverslip with rubber cement using a syringe with a 19-g needle. Make sure the coverslip is completely sealed.

9. Incubate overnight at 37 °C in a humidified chambers.

Recipes for reagents
1. 20 × hybridization salts
   Important: dissolve in the following order:
   (a) 0.1 M EDTA—18.61 g
   (b) 0.1 M Pipes—17.12 g
   (c) 3.0 M NaCl—87.66 g

Protocol 27. Continued

  (d) Bring up to 500 mls with ddH$_2$O
  (e) pH 6.8, autoclave, store at room temperature
2. 50 × Denhardt's solution
  (a) BSA—5 g
  (b) Polyvinyl pyrollidone—5 g
  (c) Ficol—5 g
  (d) Bring up to 500 ml in ddH$_2$O
  (e) Aliquot and store at −20 °C

Protocol 28. *In situ* hybridization method: day 2

Reagents needed
- 20 × SSC(see recipe)—need diluted to 2 × and 0.1 ×
- Kodak NTB2 emulsion (see recipe at the foot of this protocol)

Procedure
1. Remove the coverslips carefully and discard as radioactive waste. The best way to remove coverslips is to start at one corner of the coverslip and carefully lift. When removing glue be sure to hold the rest of the coverslip down so that the section does not peel off.
2. Dip the slides briefly in a disposable 50-ml tube containing 2 × SSC. This gets rid of the majority of radioactive liquid mix. (Discard this as radioactive waste).
3. Soak slides twice in 2 × SSC for 15 min each at room temperature.
4. Soak slides twice in 0.1 × SSC for 15 min each at room temperature.
5. [If the sections are starting to come off the slide, omit this step and proceed to step (f)]. Wash the slides for 10min in 0.1 × SSC at 60 °C. The timing of this step is crucial since it is such a stringent step in the washing process.
6. Wash once for 15 min in 0.1 × SSC at 37 °C.
7. A final wash of 10 min in 2 × SSC is performed at room temperature.

Dipping in emulsion
1. The emulsion is stored at 4 °C after it is made up. The emulsion becomes a solid at this temperature, therefore it is necessary to heat the emulsion for approximately 45 min at 42 °C.
2. The emulsion is extremely light sensitive so all of the procedures are performed in the dark.
3. Before dipping the experimental slides, use a blank test slide to determine if

Protocol 28. *Continued*

the emulsion has reached the correct viscosity. Dip the test slide vertically up and down approximately three times. (Try to keep from moving the slide in such a way as to form bubbles in the emulsion). Go outside to the light and check for three things:

(a) Make sure the colour of the emulsion is a yellowish–ivory colour. If there is any pinkish hue to the emulsion, this means it has been exposed to light—DO NOT USE THIS EMULSION.

(b) Make sure the entire slide is completely and evenly covered with emulsion.

(c) Make sure there are no bubbles in the emulsion on the slide. Bubbles will cause incomplete coverage of the emulsion on the slide.

If the emulsion looks good proceed to dip your experimental slides. After the slides are dipped, place them in a metal rack so that they can dry. The rack *must* be stored in a light-proof box or drawer.

4. The slides should be dry in 1–2 h and can then be transferred to a light-proof box. The box can also be covered with aluminium foil to ensure maximum light-proofing. The slides are then kept at 4 °C for 4 days.

Recipes for reagents

1. 20 × SSC
    (a) NaCl—173.3 g
    (b) Na citrate—88.2 g
    (c) Dissolve in 800 ml of H_2O and adjust pH to 7.0.
    (d) Bring up the volume to 1 litre and aliquot.
    (e) Autoclave to sterilize.

2. Kodak NTB2 Emulsion (can be done with red light on only)
    (a) Thaw the stock bottle of emulsion at 42 °C for 45 min. (The stock bottle contains 118 ml).
    (b) When completely thawed, add the emulsion to a beaker containing 118 ml of 0.66 M ammonium acetate. Stir very gently to prevent bubble formation.
    (c) Aliquot the emulsion in 50-ml tubes (wrapped with black tape), and store at 4 °C.

The emulsion in each tube can be used 4–5 times before the level becomes two low to completely cover the slide. At this point combine two tubes together. The emulsion can be used until the expiration date, which is printed on the suppliers box.

Techniques for double-labelling

Protocol 29. In situ hybridization method: day 6

Reagents needed
- Kodak D-19 developer (recipe on package)
- Kodak fixer (recipe on package)
- Meyers haematoxylin
- 0.5% Lithium carbonate

Procedure
1. Take the slides from the cold room and let them stand at room temperature for approx 30 min. (Keep in the dark!). At the same time place the bottles of developer and fixer on ice. The ideal temperature range for the developer and fixer is between 15–19 °C.
2. After the slides and solutions have come to the correct temperatures, the developing and fixing occurs as follows (in complete dark):
    (a) Place the slides in developer for 2 min.
    (b) Rinse in ddH$_2$O for 10 sec (this should also be cooled to 15–19 °C).
    (c) Place slides in fixer for 5 min.
3. At this point you can turn on the light. If the slides are black or partially back it means that light damage has occurred. Rinse the slides for 3 min in tap water.
4. Counterstain with Meyers haematoxylin for 2 min (if stain is older it may take a little longer).
5. Rinse in ddH$_2$O, then in tap H$_2$O and finally a brief dip in 0.5% Li carbonate. Let the slides rinse in tap water for a few minutes after the lithium carbonate. The lithium carbonate enhances the blue background staining.
6. Dehydrate the slides with ethanol: 30%, 50%, 70%, 95% and 100% and allow to air-dry.
7. Mount coverslips on to the slides with DPX and look at the slides under a microscope.

31. Immunocytochemistry—in situ double-labelling technique

This technique combines immunocytochemistry and *in situ* hybridization to achieve simultaneous detection of both proteins and nucleic acids. This technique is especially useful in determining specific cell types that are infected. The immunocytochemistry can be used to stain a cell surface marker, while the *in situ* hybridization can detect nucleic acids from a specific infectious agent.

The immunocytochemistry must be performed first. The emulsion step during the *in situ* procedure would block any accessibility of antibody to cellular antigens.

The procedure for immunocytochemistry is performed exactly as outlined in the section on single-labelled immunocytochemistry. DAB is the best detection system to use since it is stable in alcohol. The brown staining properties of DAB provide a nice contrast to the black grains of the *in situ* hybridization.

The *in situ* hybridization procedure directly follows the immunocytochemistry. After the slides have been immersed in PBS to stop the colour reaction, they are treated with proteinase K according to the procedure outlined in step(h) in the pretreatment section of the *in situ* protocol. Once the slides are treated with proteinase K, dehydrate them in graded EtOH washes (30–100%). After the slides are dehydrated, continue with the hybridization and follow the rest of the *in situ* protocol.

References

89. Wiley, C. A., Schrier, R. D., Nelson, J. A., Lampert, P. W., and Oldstone, M. B. A. (1986). *Proc. Natl. Acad. Sci. USA*, **83**, 7089.
90. Nelson, J. A., Reynolds-Kohler, C., Oldstone, M. B. A., and Wiley, C. A. (1988). *Virology*, **165**, 286.
91. Wiley, C. A. and Nelson, J. (1988). *Am. J. Pathol.* **133**, 73.
92. Wiley, C. A., Oldstone, M. B. A., and Nelson, J. A. (1987). *J. Neuropathol. Exp. Neurol.*, **46**, 348.
93. Nelson, J. A., Wiley, C. A., Reynolds-Kohler, C., Reese, C. E., Margaretten, W., and Levy, J. A. (1988). *Lancet*, **i**, 259.
94. McLean, W. and Nakane, P. K. (1974). *J. Histochem. Cytochem.*, **22**, 1077.
95. Singer, R. H., Lawrence, J. B., and Villnave, C. (1986). *BioTechniques*, **4**, 230.
96. Haase, A., Brahic, M., and Stewring, L. (1984). *Methods in Virol.*, **7**, 189.
97. Wood, G. S. and Wainke, R. (1981). *J. Histochem. Cytochem.*, **29**, 1196.
98. Navitoku, W. Y. and Taylor, C. R. (1982). *J. Histochem. Cytochem.*, **30**, 253.
99. Feinberg, A. P. and Vogelstein (1983). *Anal. Biochem.*, **132**, 6.

3

Detection of immune complexes and antibody mediated target cell lysis

MICHAEL B. A. OLDSTONE

1. Immune complexes in body fluids during viral infection

Experience with a large number of viral infections in animals and humans indicates that immune complexes form and can be detected both in the circulation, as well as in target tissues, where they localize. Presumably, the interaction of viral antibody with virus or viral antigens in the circulation during acute and chronic virus infection not only serves as a marker for the presence of virus but also produces immune complexes that commonly cause nephritis and arteritis. Immune complex deposits are the source of tissue damage that is immunopathologically similar in humans and animals and is likely to be responsible for several of the signs and symptoms that accompany acute viral infections.

1.1 Test to detect immune complexes

Evidence for the presence of immune complexes lies in demonstrating circulating complexes and in showing localization of antigen, host immunoglobulin, and complement components at the site of tissue injury. Usually, non-antigen-specific tests are performed to determine the presence of such complexes initially. Of the many techniques developed to demonstrate and quantitate circulating immune complexes we have found two of general usefulness, the Clq-binding test and the Raji cell test. Both of these assays are based mainly on the interaction of immune complexes with complement components or complement receptors. Although both assays frequently give comparable results, the Clq test may be more sensitive in detecting complexes formed in antigen excess, whereas the Raji test may have enhanced sensitivity for detecting immune complexes formed in antibody excess. Hence, the tests probably complement one another. The disadvantages of these tests are that they only measure complement-bound immune complexes and, although they measure activity occurring during viral infections, they are in themselves not virus- or antigen-specific. Additionally, during virus infections, B cells may be polyclonally activated with production of autoantibodies. If anti-lymphocyte antibodies are made the Raji assay can give a

Immune complexes

false result. For that reason, in experimental studies we have favoured and used the Clq assay. Additionally, the Clq assay allows isolation and concentration of the complex. Techniques to measure antigen specificity in the circulating complex are used. The reactants in the complex can then be isolated and segregated by treatment with SDS and polyacrylamide electrophoresis and identified using specific probes against viral proteins (see Chapter 2).

The principle of the Clq binding assay is to measure the amount of radiolabelled Clq that binds to macromolecular substances in a sample of fluid. Unbound Clq is separated from Clq bound to complexes by precipitation with polyethylene glycol (PEG).

1.2 Purification of Clq
Clq purified from human sera can be used for human or murine studies.

Protocol 1. Purification of Clq

1. Obtain serum 2 h after clot formation and centrifuge it at 30 000 g for 30 min at 4 °C so that any lipid layer formed at the top of the tube can be removed.
2. Mix 40 ml of serum with 10 ml of 0.1 M EDTA, pH 7.5, for 10 min at 37 °C and immediately place the mixture in an ice bath. Adjust the pH to 7.5. All subsequent steps are done with ice-cold reagents and in an ice bath.
3. Slowly add 200 ml of 0.005 M EDTA, pH 7.5, to the serum–EDTA mixture while it is gently stirred with a glass rod. Leave the final mixture in the ice bath for 1 h and gently stir it every 15 min.
4. Centrifuge at 12 000 g for 30 min at 4 °C, and recover the precipitate formed. Gently resuspend it in 80 ml of 0.022 M EDTA. It is important that the conductivity of the 0.022 M EDTA corresponds to that of 0.04 M NaCl.
5. Recentrifuge and *resuspend* the precipitate in 0.022 M EDTA.
6. After a *repeat* centrifugation, collect the precipitate and dissolve it in approximately 10 ml of 0.75 M NaCl, 0.01 M EDTA, pH 5.
7. Hold the solution at 4 °C overnight then centrifuge it at 30 000 g for 30 min at 4 °C.
8. Collect the supernatant fluid and dialyse it for 4 h against 0.067 M EDTA, pH 5 (the conductivity of this solution should correspond to that of 0.078 NaCl). A precipitate should form which is recovered after centrifugation at 12 000 g for 30 min at 4 °C.
9. Wash this precipitate twice in 0.067 M EDTA, pH 5, and gently dissolve it in 3 ml of 0.3 M NaCl, 0.01 M EDTA, pH 7.5, for 2 h at 4 °C.
10. Centrifuge the solution at 30 000 g for 30 min at 4 °C to remove aggregates, after which the supernatant fluid contains purified non-aggregated Clq. A yield of 1–2 mg from 40 ml of serum is expected.

Protocol 1. *Continued*

11. Check the purity immunoelectrophoretically using 1% agarose in buffer containing 10 mM EDTA and test against antiserum to total human serum and antiserum to Clq.
12. Divide the purified Clq into portions containing approximately 250 µg of Clq each and store at $-70\,°C$ until required.

These preparations can be used as long as 6–12 months after purification.

1.3 Radiolabelling of Clq

Purified Clq is labelled with ^{125}I (~ 350 mCi/ml) by using lactoperoxidase iodination.

Protocol 2. Radiolabelling of Clq

1. Dissolve 250 µg of Clq in 500 λ of 0.3 M NaCl and 0.01 M EDTA, pH 7.5.
2. Add to this mixture 5 λ of ^{125}I (~ 500 µCi in NaOH), 5 λ of NaI [0.006 mg/ml in Veronal-buffered saline (VBS), 0.01 M sodium barbital, 0.5 M NaCl, 1 mM $CaCl_2$, 1 mM $MnCl_2$, pH 8]; 5 λ of lactoperoxidase (1 mg/ml in VBS); 5 λ of H_2O_2 (3×10^{-3}% in VBS).
3. Gently agitate the mixed reagents and then incubate them for 15 min in an ice bath.
4. Stop the reaction by the addition of 10 λ of NaI in VBS (6 mg/ml), 10 λ of NaN_3 in VBS (0.03 mg/ml), 2 ml of VBS with gelatin 0.1% or bovine serum albumin (BSA) carrier. The radioactive uptake is generally between 50–80%, and the ^{125}I Clq has a specific activity of more than 1 µC/µg (range 1–1.6 µC/µg.
5. Add radiolabelled Clq to a Con-A Sepharose column 1×14 cm. Add 1-ml volumes of 0.5 M NaCl in VBS; collect the eluted fractions and determine radioactive counts. This procedure removes fragments of Clq or possible immunoglobulin contamination in the Clq preparation.
6. After radioactive counts have fallen to baseline levels, elute the labelled Clq from the Con-A Sepharose column by adding 1 ml volumes of 0.5 M NaCl in VBS containing 10% α-methyl-D mannoside.
7. Monitor individual fractions in a gamma counter, pool and stabilize material at the peak by the addition of BSA to give a final concentration of 0.1% BSA in 0.15 M NaCl in VBS buffer.
8. Dialyse against VBS buffer in the cold, then ^{125}I Clq can be stored at $-75\,°C$. Usually >96% of ^{125}I Clq is precipitated by TCA.

Once iodinated, the radiolabelled Clq is generally not used for more than 14 days.

Protocol 3. Clq assay

1. Centrifuge ^{125}I Clq at 10 000 r.p.m. for 40 min before use in order to remove aggregates.
2. Dilute the deaggregated ^{125}I Clq in 1% BSA in 0.5 M NaCl in VBS so that approximately 4 000 000 counts of ^{125}I Clq are present in 1 ml.
3. Coat plastic test tubes, 12 × 75 mm, with 0.1% gelatin in VBS, invert them and allow to dry overnight.
4. Place 25 λ of the serum sample to be tested in labelled test tubes with 50 λ of 0.2 M EDTA, pH 7.5.
5. For the standard curve, place 25 λ of aggregated human gamma globulin (see below) in labelled test tubes with 25 λ from a control pool of serum (negative for Clq-binding activity, heated at 56 °C for 30 min) and 25 λ of 0.2 M EDTA, pH 7.5.
6. Incubate for 30 min at 37 °C, then transfer the tubes immediately to an ice bath and add 25 λ of ^{125}I Clq ($\sim$100 000 c.p.m.).
7. Immediately thereafter, gently add to each tube 1 ml of PEG, 3% in borate buffer (0.1 M boric acid, 0.025 M sodium tetraborate, 0.075 M NaCl, pH 8.2–8.4).
8. Do not agitate tubes but leave them to stand in the ice bath. After 1 h of incubation, centrifuge the tubes for 40 min at 2200 r.p.m.; drain the supernatant fluid completely, and measure the radioactivity in the precipitates.

Always run samples in triplicate, and run concurrent controls such as a known negative control (pooled human serum without Clq-binding activity), 25 λ of ^{125}I Clq mixed with 75 λ of FCS and 1 ml of 20% TCA, add standard amounts of aggregated human γ globulin. The soluble heat aggregated human γ globulin is formed from a solution of 10 mg Cohn fraction II per ml, NaCl 0.9%, heated at 63 °C for 20 min, and centrifuged at 10 000 r.p.m. for 10 min. Standard mixtures of aggregated human γ globulin are used which contain 10-fold dilutions from 10 mg/ml down to 1 μg/ml.

1.4 Standard curve and interpretation

The results are expressed as the percentage of TCA-precipitable Clq radioactive counts precipitated by PEG. In our experience, control human or mouse serum precipitates 5%$\pm$2 (mean$\pm$1 SD) of added TCA-precipitable ^{125}I Clq. In comparison, the various standard solutions of aggregated human γ globulin mixed with heat inactivated serum at concentrations of 10 mg, 1 mg, 100 μg, 10 μg, and 1 μg bind approximately 100%, 67%, 28%, 18% and 10% of the TCA-precipitable, radiolabelled Clq counts that are offered. These data are most meaningfully expressed as the percentage of TCA-precipitable Clq bound, rather than as equivalents of aggregated human γ globulin, since the aggregates vary among different preparations and may not reflect the size and nature of the immune complexes detected. When the presence of immune complexes is tested in mouse serum, 100 λ (instead of 50 λ) of 0.2 M EDTA, pH 7.5, is used. Experience in several laboratories has indicated that freezing and thawing the serum sample 3–5 times usually does not alter the Clq-binding efficiency. However, it is best to disperse serum being tested in small volumes to avoid repeated freezing and thawing. Some complexes containing only IgA have been reported as not binding Clq.

2. Infectious virus–antibody immune complexes in body fluid

It is clear from the study of several persistent virus infections (lymphocytic choriomeningitis virus, lactic dehydrogenase, aleutian disease of mink, equine infectious anaemia, etc.) that infectious viruses can travel in sera complexes with antibody and complement. In terms of pathogenesis this not only results in deposition of immune complexes in tissues like glomeruli, arteries and choroid plexi with potentially resultant disease but also offers the virus an opportunity to enter cells containing Ig, Fc or complement receptors. Further, in a number of virus systems, most notably reovirus and dengue it is clear that both enhanced virus uptakes by the complex over the virus itself and/or different locations in unique cell compartments of virus alone versus virus–antibody complex can occur. Selection with virus–antibody complex apparently favours viral persistence.

2.1 Detection of infectious virus–antibody immune complexes

Infectious virus–antibody complexes are detected by placing aliquots of sera (25–100 λ) in four separate Eppendorf tubes. The first tube serves as the positive control and receives equal aliquots (volume) of non-specific (non-immune–viral) reagents. The second tube receives antibody to the host immunoglobulin (Ig). Sufficient antibody is added to remove all the host's Ig (usually IgG). For monitoring a trace amount of ^{125}I-labelled IgG is added. The antibody to host IgG is added until ^{125}I counts are removed from the serum fluid in the third tube and similarly treated but with antibody to host C3 (3rd component of complement). The sample in the last tube is similarly treated with antibody to host albumin (negative control). For immunoprecipitation high titred affinity purified antibodies are preferred. IgM or IgA containing virus complexes may occur and are assayed for in a similar manner. After removal of complexes by immunoprecipitation and centrifugation (10 000 g for 5–10 min), the remaining supernatant is analysed for numbers of viral plaque-forming units.

3. Tissue deposited virus–antibody immune complexes

Once circulating immune complexes deposit, they can be recognized on the basis of their immunopathological patterns. They form a discontinuous granular deposit of host immunoglobulin, specific viral antigen(s) and complement components in the absence of deposits of other (trapped) serum proteins (i.e. albumin, fibrinogen) in the lesion. Such complexes most often deposit in glomeruli, arteries and the choroid plexus of the brain.

3.1 Detection of tissue bound virus–antibody immune complexes

To detect complexes a 4-μm section is made from snap-frozen fresh tissue (preferred method) or paraffin-embedded tissue. A glass slide containing the tissue section is fixed in ether–alcohol (volume 1:1, 10 min) and 95% ethanol (20 min), washed in PBS (5 min) and stained with the appropriate immunochemical reagents. Antibody to host Ig (IgG), or to host C3 should be positive while antibodies to host albumin or host fibrinogen should be negative. With fresh tissue the antibody is conjugated to a fluorochrome dye and material studied under a fluorescent microscope. For paraffin sections a peroxidase-labelled antibody is often used. To detect viral antigen, it is often necessary to first treat the tissue with solutions that dissociate antibody–antigen complexes (i.e. low pH, low molar citrate or glycine buffers). The buffer condition needs to be adapted to the particular antigen–antibody complex system. Thereafter the section is treated with antibody to the appropriate viral antigen.

4. Antibody mediated injury of virus infected cells

4.1 Antibody and complement mediated injury

Antibodies can injure virus infected cells by two distinct pathways. By the first, antibody and complement bind to a virus infected cell and cause its lysis. To assay for this possibility serum taken during virus infection is added to ^{51}Cr-labelled virus infected target cells. When such sera contain cytolytic antiviral-antibody and a functioning complement source the target cell will be lysed with the degree of ^{51}Cr released quantitating the degree of injury. To accomplish this assay, 100 λ of sera is untreated, an additional 100 λ of sera is treated for 30 min at 56 °C [to inactivate the alternative pathway (50 °C) and classical pathway (56 °C) of complement], a third 100 λ of sera is incubated for 30 min at 56 °C and 50 λ of alternate complement source (usually guinea pig) is added. Negative controls consist of 100 λ of pre-infection serum and 100 λ of pre-infection serum with 50 λ of the alternate complement source. These reagents are added to 2×10^4 ^{51}Cr-labelled virus infected and uninfected target cells. Incubation is at 37 °C for 1 h and carried out in the presence of veronal complement buffer (see 3, above).

4.2 Antibody-dependent cell mediated cytotoxicity

Antibody in collaboration with non-immune lymphocytes also lyses virus infected target cells. Here the Fab parts of the antibody molecule bind specifically to the viral antigens expressed on the cells' surface, leaving the Fc piece to react with any lytic cell bearing an Fc receptor. Various dilutions of serum (1/5 to 1/1000), previously heated at 56 °C for 30 min, are added to virus infected target cells and allowed to incubate for 5–10 min. Thereafter, PBL harvested from normal non-immune donors are added, and the release of ^{51}Cr at the end of 6 h is monitored. Controls consist of using a serum source devoid of antibodies. Handling of cultures and calculations are similar to that described above. The advantage of this assay is its great sensitivity with the ability to detect, in some systems, antibody activity in serum diluted 1×10^{-5}. In addition, this assay is 20- to 50-fold more sensitive in detecting antibodies than conventional immunofluorescent or complement-associated tests.

4

Detection, generation, and use of cytotoxic T lymphocyte (CTL) clones

4A. DETECTION OF CTL ACTIVITY
MICHAEL B. A. OLDSTONE

1. Introduction

Cytotoxic T lymphocytes (CTL) recognize an infected target cell in a virus-specific major histocompatibility antigen (MHC) restricted manner (*Figure 1*).

1.1 Selection of target cells

Target cells are selected on the basis of containing the correct MHC glycoproteins, permissibility for the infecting virus, ease of manipulation and carrying in culture, stability of the uninfected cell in releasing low amounts of internalized radiolabelled ^{51}Chromium, and ability of the cell to be lysed by CTL during which maximal amounts of ^{51}Cr are released. Established MHC typed murine lines most often used because they fit the above criteria are listed in *Table 1*.

When study of other MHC haplotypes or of MHC recombinants (i.e. $K^bI^bD^b$, $K^bI^bD^d$, $K^dI^dD^b$, etc.) is desired, mouse embryo fibroblasts of the preferred MHC background are made. Such lines have a finite life span, so a large masterseed is frequently made at the third passage and aliquots of cells frozen until use for

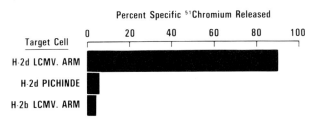

Figure 1. CTLs are generated in H-2^d mice using LCMV Armstrong (ARM) virus. These CTL efficiently and effectively kill H-2^d LCMV ARM but not H-2^b LCMV ARM infected targets (*H-2 (MHC) restriction*). Such CTL kill only H-2^d targets infected with LCMV ARM, the immunizing virus but not Pichinde virus (*virus specificity*).

Table 1. MHC typed murine lines

MHC locus			Cell line used
K	I	D	
k	k	k	L-929
d	d	d	Balb C17, p-815
b	b	b	MC57

further study. Some of these lines have been transformed with any of a variety of transforming agents so that they can be used as continuous lines. Additionally, a limited number of cell lines are available that express a transfected specific MHC gene.

When studying CTL responses in humans, the availability of cells is more limited. Appropriate targets are generated most often in either of two ways. First, skin biopsy is obtained from a MHC typed person (HLA A, B, C, D locus typed) and the firbroblast line obtained. A battery of lines containing different HLA haplotypes is collected. Because it takes several weeks to grow out such fibroblast lines before sufficient amounts are available for assay, this approach is most often used to study the generation of CTL in vaccinated healthy individuals, or CTL responses of healthy individuals to viruses they have previously encountered. A second method allows the above studies to be performed in addition to the analysis of CTL activity of ongoing viral infection. The individual's B lymphocytes are obtained from peripheral blood, labelled with ^{51}Cr and transfected with a suitable vector capable of expressing the viral protein under study (see 4.1a). These cells are then mixed with the same individual's T lymphocytes.

1.2 Preparation of lymphocytes from blood

Separation of lymphocytes from peripheral blood can be achieved as shown in *Protocol 1*.

Protocol 1. Preparation of lymphocytes from blood

1. Transfer aseptically up to 50 ml of blood obtained from the anticubital vein to 50 ml plastic tubes containing heparin (20 units of heparin with no preservative/ml blood). All the following manipulations are done at room temperature except when otherwise specified.
2. Dilute the venous blood in heparin, 1 part to 2 parts, with PBS and layer on top of Ficoll–Paque sterile solution (Pharmacia Fine Chemicals, Piscataway, NJ).
3. To a 50-ml glass or plastic tube, add 15 ml of Ficoll–Paque and carefully layer 30 ml of diluted blood over it.

Protocol 1. *Continued*

4. Centrifuge the mixture at 800 g for 25 min or 400 g for 30–40 min, after which the interface layer should be carefully collected by using a Pasteur pipette, and transferred to a sterile test tube.
5. Wash the cells in the interface (usually 94% or more mononuclear cells) twice in a balanced salt solution (MEM) and centrifuge at 700 g for 10 min.
6. After the final wash, resuspend the mononuclear pellet in RPMI 1640 medium containing 20% fetal calf serum (FCS).

There is little difference between using human AB-type serum or FCS in the cytotoxic assay, as long as the serum source does not contain antibodies to the virus being assayed. Mononuclear cells are counted; their viability is checked by the addition of trypan blue or eosin, and their concentration is adjusted to 5×10^6 mononuclear cells/ml of RPMI 1640 medium plus 20% FCS. The majority of monocytes can be removed by culturing mononuclear cells on glass or plastic surfaces and collecting the non-adherent cells after 1 h at 37 °C in 5% CO_2 or by lysing cells with any one of several commercially available anti-macrophage monoclonal antibodies and complement source. For special studies, the peripheral blood lymphocytes (PBL) can be divided into B and T subpopulations based on the presence or absence of specific receptor molecules, and use of monoclonal antibodies directed against these molecules either with complement (to lysis), antibodies coated to a plastic dish (to PAN) or with a fluorochrome dye (fluorescent activator cell sorting).

1.3 ^{51}Chromium-labelling of cells

Optimal results are obtained with established murine lines while lesser results are achieved with fibroblast or B lymphocyte targets. The data given reflects the use of these established cell lines. Usually 100 λ containing 0.1 mCi (3.5 μg Cr per ml) of ^{51}Cr is given to 1×10^6 cells and 1×10^4 ^{51}Cr-labelled target cells are placed in the well for assay. The maximal ^{51}Cr release as determined by treatment with 1% NP-40 is 80–90% of ^{51}Cr uptake and ranges from 5000 to 15 000 C.N.T./min. Most assays are run for 5–6 h. By then the spontaneous release is in the range of 500 to <1500 CNT/min given a sufficient window for accurate measurements. Free ^{51}Cr is taken up by the cell where it binds to a cytoplasmic protein. Protein bound ^{51}Cr is not taken up by bystander cells. Some cells take up ^{51}Cr poorly, even though culture conditions and temperature is optimized. Cells that change from their usual characteristics and now spontaneously release enhanced amounts of ^{51}Cr may be contaminated by bacteria or mycoplasma.

The usual procedure for uptake of ^{51}Cr into cells is achieved by mixing target cells with ^{51}Cr (200 μCi/2×10^6 cells) for 45 min at 37 °C at 5% CO_2. Thereafter cells are washed three times, and 1–2×10^4 ^{51}Cr-labelled target cells are dispensed into each well of a 6-mm flat bottomed microtest plate.

1.4 Effector CTL

Effector CTL (not cloned) have a low precursor frequency during primary infection that increases during secondary immunization. Hence, CTL from spleen, lymph nodes or peripheral blood are used most often at effector to target ratios of 50:1. In systems of low CTL frequencies, a ratio of 100:1 can be employed. Higher numbers of CTL often lead to unacceptable background (i.e. high spontaneous release of uninfected targets) and/or a prozone effect. At least two and preferably three effector to target ratios are used during the assay with ratios most often being 50:1, 25:1, and 12.5:1. Washed effector cells are suspended in media with 10% FCS prior to being added to targets.

Effector cells are added to virus infected and uninfected targets labelled with ^{51}Cr. The optimal assay time is 5–6 h although this time can be extended up to 18 h when CTL frequency is low.

1.5 Evaluation and calculation of CTL assay

Each test should be performed at least in triplicate. Variability between triplicate samples rarely exceeds 10% and averages 4% of the total sample count. Controls recording both spontaneous ^{51}Cr release and maximal experimental release should be included in each experimental run. Maximal release is determined by the addition of 150 λ of 1% nonidet (NP-40) at the end of the incubation period; this usually releases 80–90% of the total radioactivity. Reactants are incubated at 37 °C in a 5% CO_2 incubator. At the end of 4–6 or 12–18 h, 100 λ cell-free samples (cell debris removed by centrifugation at 400 g for 10 min using a microtitre holder attachment) are removed and placed into tubes, and the remaining radioactivity is counted in a gamma counter.

The percentage of ^{51}Cr released is calculated according to the following formula:

$$\frac{E - S}{Max - S} \times 100$$

Where E = ^{51}Cr released from infected target cells in the presence of CTL, S = ^{51}Cr released from infected target cells in the presence of medium alone, and Max = ^{51}Cr released upon addition of water and NP-40. ^{51}Cr released is calculated separately in each condition for both infected and uninfected cells, the specific release induced by CTL is subtracted from the values obtained with non-infected target cells. Caution in interpreting results should be used when the spontaneous release exceeds 35%. The assay should be run under conditions (m.o.i., length of test) where the spontaneous release of ^{51}Cr from uninfected and virus infected cells are approximately the same. Experiments should contain controls testing for MHC restriction and virus specificity (*Figure 1*). There are reports that normal (non-sensitized, non-immune) PBL may have lytic activity against cultured cell lines (natural killer cell lysis). In our experience, lysis frequency is on occasion greater against uninfected than infected targets and usually close to the values for spontaneous release of ^{51}Cr from target cells in the absence of PBL and hence does not interfere with the above assay.

4B. GENERATION AND USE OF CTL CLONES
HANNA LEWICKI

2. Primary CTL response

Primary CTL response is generated in mice and other experimental animals by inoculating 1×10^4–1×10^5 p.f.u. of virus intraperitoneally. Depending on the virus (family or strain within a family) and host permissiveness, peak CTL titres are usually found in the spleen 6.5–9 days after inoculation. As a rough guide, splenic enlargement, i.e. *in vivo* CTL lymphocyte proliferation and expansion, often correlates directly with CTL activity. Spleens are aseptically removed, kept on ice and kept wet with cold media containing 10% FCS. Single cell suspension is made by finally teasing the spleen with surgical instruments or pushing it through a 100-μm mesh screen to remove aggregates. Cells are washed thoroughly in media containing 7% FCS with alternating centrifugations at 800 r.p.m. for 7 min. Red cells are removed by the addition of 0.83% NH$_4$Cl. NH$_4$Cl is warmed at 37 °C and added to a pellet of packed cells (1 ml of 0.83% NH$_4$Cl per spleen equivalent). After gently pipetting back and forth twice, media is added [2 vol of media to 1 vol (NH$_4$Cl+spleen)] and the resultant mixture recentrifuged at 800 r.p.m. for 7 min. Supernatant is removed and cells resuspended in 10 ml of media. Effector cells are counted and placed at a concentration of 1–2 $\times 10^6$ cells/ml of media containing 7% FCS.

Secondary CTL response is generated from mice usually 30–60 days after being primed (see above) with 1–2 $\times 10^5$ p.f.u. of virus. Spleen or lymph node lymphocytes are harvested as described above and added to syngeneic macrophages infected with the virus of interest. Two $\times 10^5$ macrophages, irradiated with 2000 rads, and infected with the appropriate virus (m.o.i. of 3) are incubated with 4×10^6 spleen cells per ml of RPMI 1640 containing 10% fetal calf serum, 5×10^{-5} M 2-mercaptoethanol, penicillin and streptomycin. After 5 days in culture, lymphocytes are harvested for biological studies, i.e. CTL assay or further manipulation in culture for the generation of CTL clones (see below).

3. CTL clones

CTL clones are made from secondary CTL cultures. Several mouse haplotypes have been successfully used to generate CTL clones. Often some haplotypes may present difficulties in going from CTL cultured line to CTL clones. In our experience these difficulties can often be overcome by changing the culture conditions and T cell growth factor concentration used and/or utilizing H-2 haplotypes on B10 backgrounds. Below we list our standard protocol for generating CTL clones which is shown in *Figure 2*.

LCMV-Arm

↓ 4–8 weeks

Spleen removed
Single cell suspension prepared
↓

2° *in vitro* stimulation
Responders: LCMV-infected Mɸ stimulators — 40:1

↓ 8–14 days

Cloned by limiting dilution in 96-well plates:

   Feeder cells: 5×10^5 syngeneic spleens/well (2000 rad)
   Responder cells: 0.3–100 cells/well, collected from 2° *in vitro* stimulation
   TCGF: 5% Con A supernatant

↓ 14–20 days

Clones expanded in 24-well plates:

   Feeder cells: 4×10^6 syngeneic spleen cells/well (2000 rad)
   Stimulated cells: 2×10^5 syngeneic LCMV-infected Mɸ's/well (2000 rad)
   TCGF: 5% Con A supernatant
↓

Clones maintained:

   Restimulated every 7 days with syngeneic LCMV-infected Mɸ's (2000 rad) and
      syngeneic feeder cells (2000 rad)
   Fed on day 4 with 5% Con A supernatant

Figure 2. The procedure used to generate CTL clones.

4. Harvesting and use of peritoneal macrophages

Three days prior to harvesting inject each mouse with 3 ml of thioglycolate solution i.p. Use a 25 gauge needle.

On the day of harvesting, proceed as detailed in *Protocol 2*.

Protocol 2. Harvesting peritoneal macrophages

1. Prepare a syringe with 6 ml of 7% FCS in RPMI medium; attach a 27 gauge needle.
2. Kill the mouse; pin it to the board.
3. Wash the mouse skin with 70% ethanol.
4. Make a horizontal cut (in the skin only) then a vertical cut up to the rib cage area.
5. Peel the skin back with forceps, check to make sure the peritoneum is unpunctured and intact.
6. Wash the exposed area with 70% ethanol.
7. Inject the 6 ml of 7% FCS RPMI via a 27 guage needle into the peritoneum.
8. Remove the needle, gently massage the peritoneum by hand for about 1 min.
9. Gently lift up the peritoneum with forceps. Using a 6 cm^3 syringe and a 23 gauge needle draw off liquid from the peritoneal cavity. Transfer the peritoneal fluid to a sterile centrifuge tube placed on ice. Peritoneal fluid should be yellow and not bloody. Use sterile tissue culture procedures (work under laminar flow hood).
10. Fill the tube with 7% FCS in RPMI medium.
11. Spin cells down at 800 r.p.m. for 8 min at either 4 °C or at room temperature.
12. Remove the supernatant.
13. Gently resuspend the cells in 7% FCS in RPMI medium and count them.
14. Bring the cells to the desired volume and concentration. Distribute cells to appropriate plates.
    96 well plates: 2×10^4 cells/well in 100 μl 7% FCS in RPMI medium.
    24 well plates: 1.5×10^5 cells/well in 0.5 ml of 7% FCS in RPMI.
15. To make thioglycolate solution: use Difco fluid thioglycolate medium dehydrated. Add 29.8 g up to 1 litre of millique water.
    (a) Boil for 10–15 min.
    (b) Aliquot into 100-ml bottles.

Generation and use of CTL clones

Protocol 2. *Continued*

  (c) Autoclave.

  (d) Store at room temperature in the dark.

Do not use immediately after preparation. For best results use ~2 weeks after preparation.

Protocol 3. Preparation of T cell growth factor (TCGF)

1. Prepare suspensions of spleen cells from 8–10 week old Lewis rats. Use 0.83% NH_4Cl to lyse the red cells. Resuspend lymphocytes in 5% RPMI medium + 10% SC (see *Protocol 2*).
2. Count the cells and resuspend them to a concentration of 5×10^6 cells/ml.
3. Put 2.5×10^8 cells in a T75 flask with 50 ml of 5% RPMI medium + 10% SC (see recipe for supplement complete at the foot of this protocol). Add ConA at a concentration of 5 µg/ml = 250 µg/T75 flask (50 ml of medium).
4. Incubate overnight in incubator 5% CO_2 at 37 °C. The next day obtain supernatant, filter (sterilization) and aliquot. Store in a freezer at −20 °C. Check each batch of TCGF by IL-2 assay.

(*SC*) *Supplement complete for mouse:*
- Hepes (Gibco)—23.83 g
- Pen/Strep (Gibco 20 ml lyophilized, 10^4 U or µg/ml)—10 bottles
- L-glutamine (Gibco)—4.0 g
- 2-Mercaptoethanol (14.3 molar strength)—70 µl
- Gentamycin, SO^4 (Irvine Scientific, 50 mg/ml)—1.000 mg

Bring this mixture to one litre with RPMI 1640 and add one litre of inactivated (30 min in a waterbath at 56 °C) fetal calf serum. Mix, filtrate, and aliquot. Use 10% SC in 5% FCS serum RPMI.

Protocol 4. Cloning CTL by limiting dilution

1. Refer to Section 3 on making CTL clones. Clones are collected on day 8–9 after initiating a secondary CTL response. Cells are added to irradiated macrophages (see below) from wells in a 24-well culture plate (4 wells should provide sufficient cells).
2. Adjust to 1×10^5 cells/ml in cloning medium (5% RPMI + 10% SC + 5% TCGF).
3. Make dilutions in cloning medium of 10^4, 10^3, 10^2, 30, 3, 1 and 0.3 cells/well.
4. Make the following plates:
   - 1 plate—6 rows 100 cells/well
       —2 rows each 10^4, 10^3, and only macrophages/well (control)

Protocol 4. *Continued*
- 1 plate each—30 cells/well
  —10 cells/well
  —3 cells/well
  —1 cells/well
  —0.3 cells/well
5. Plate cells in 96 wells that already contain irradiated (2000 rads) macrophages (called macrophage plates).

  Macrophage plates can be prepared in advance. Place 2×10^4 peritoneal macrophages in each 96-well plate. Add 100 μl of 7% FCS in RPMI medium. Add feeder cells: 100 μl/well (see below).

Feeder cells
(a) Harvest spleen cells from syngeneic mice.
(b) Remove red blood cells.
(c) Treat splenic lymphocyte cells with 2000 rads.
(d) Adjust cell concentration to 1×10^7/ml in cloning medium.
(e) Add virus at a concentration of 10^3 p.f.u./ml of feeder cells, add in a volume of 100 μl/well in 96-well plate prior to adding responder cells.

Protocol 5. Carrying and expanding CTL clones
1. Feed every 3–4 days with cloning medium.
2. After approximately 1 week add 5×10^5 irradiated feeder cells/well with virus; add this every week.
3. You should be able to screen positive cells by 3–4 weeks.
4. Discard plates after 4 weeks if no growth has occurred.
5. Take clones from a plate where less than 20% of the cells are positive.
6. Expand from one 96-well to one 24-well plate. Plates should contain:
   1×10^5 macrophages, irradiated and infected with virus.
   5×10^6 splenic lymphoid cells receiving 2000 rads.
7. Pass cells again when ready; feed with cloning medium.

Protocol 6. Freezing and storing CTL clones
1. Preparation of freezing solution: 5% DMSO in fetal calf serum.
2. The day before freezing, change the medium.
3. For best results freeze cells 3–4 days after stimulation.
4. Freeze $2–3 \times 10^6$ cells/ml.

Protocol 6. *Continued*
5. To re-use frozen CTL clones:
   (a) Thaw cells to 37 °C
   (b) Plate on macrophage feeder layers with TCGF-media.

5. Use of CTL clones *in vivo*

5.1 Adoptive transfer of cloned CTL

The CTL clones to be used for adoptive transfer are harvested 5 days after their last passage *in vitro*. Cells are washed once in serum-free RPMI 1640, gently dissociated with 0.5% trypsin (Irvine Scientific), washed three times with RPMI 1640 (containing 7% FCS, 1 mM glutamine, 1×10^4 U/ml penicillin and 10 mg/ml streptomycin), before intravenous, intraperitoneal or intracerebral inoculation into recipient mice. The amount of CTL transferred varies according to the experiment proposed. Usually $5 \times 10^5 - 1 \times 10^7$ or 10^8 cells are transferred. A spleen contained about 1×10^8 lymphocytes. For intracerebral transfer 5×10^2 cells can cause effective immunopathology although 1×10^7 cells can be transferred.

6. Ablation of the host's immune response

To determine the ability of cloned CTL to induce virus disease during acute infection, recipient mice are immunosuppressed with 750 rads irradiation from a ^{137}Cs source (to ablate their immune response) and inoculated with virus to initiate infection. At varying times thereafter, depending on the virus, but usually 3–5 days after infection, recipients are adoptively transferred with cloned CTL in 50 μl of medium.

4C. USE *IN VITRO* TO MAP CTL EPITOPES
J. LINDSAY WHITTON and ANTOINETTE TISHON

7. Introduction

The cellular immune response to viral infection characteristically is polyclonal in nature: that is, the bulk CTL response detected by *in vitro* assay against virus infected cells reflects the cumulative activity of several distinct cell populations, each directed against a different viral epitope. The generation of monospecific CTL lines, or clones, and the *in vitro* assay (^{51}Cr release) have been described in Chapter 4B. Here we shall detail the further characterization of CTL clones. What virus protein does each clone see? Having determined this, what particular region of the protein (i.e. epitope) is recognized, in association with MHC, by the individual T cell receptors (TCR)?

8. Identification of virus target protein

For many years the precise target proteins for CTL remained obscure. The advent of monospecific antibody preparations brought with it the possibility that these agents could be used to block CTL activity, thereby identifying (albeit by default) the target antigen. Although in principle the hypothesis appeared sound, as antibody to class I MHC glycoprotein (the second component of the target complex) could markedly diminish CTL recognition, in practice most virus-specific antibodies had no effect whatever. We now understand that the nature of virus antigen recognized by the two molecules (antibody or TCR) is very different.

8.1 Use of reassortant viruses

Successful elucidation of viral target proteins thus came to rely instead on techniques which allowed the viral proteins to be expressed in isolation from their normal consorts. Such studies were particularly readily undertaken using viruses with segmented genomes, where reassortants could be generated. Such studies on, for example, influenza virus (an 8-segment RNA virus) and lymphocytic choriomeningitis virus (LCMV: a bisegmented RNA virus) provided provisional answers to the question.

8.2 Use of recombinant DNA techniques

Recent studies have used recombinant DNA technology to express individual virus proteins. Two general routes have been followed. One, exemplified by the initial studies on influenza virus, and subsequent analysis of the immunogenicity of SV40 T antigen, used gene transfection techniques to isolate stably transformed cell lines which constitutively expressed the protein of interest. The gene of interest is cloned into a plasmid expression vector containing all of the necessary transcriptional control signals, as well as a selectable marker gene such as that encoding resistance to the anti-metabolite G418. This DNA is then introduced into a cell line *in vitro*. At best, only 1 cell in 10^1–10^2 will take up and express such DNA; and the vast majority of these will fail to retain it. Only a tiny minority of cells, perhaps 1 in 10^5, will retain and express the DNA (which recombines into the host genome). Such a cell is said to be stably transformed. This cell is selected from the background cells (which fail to express the introduced sequences) by incubation with G418; only the stably transformed cells will survive and multiply. As a general rule, cells expressing the marker gene (e.g. G418 resistance) will also express the (unselected) gene of interest. The end result is a line in which all cells express the desired product. The advantages of this technique are:

- it is well characterized;
- it involves relatively straightforward cloning procedures;

- a potentially valuable reagent is generated in which the target antigen is continually expressed in the absence of cytopathology.

The major drawbacks are that
- the protein is expressed in a single MHC background; expression on other backgrounds requires the isolation of new stably transformed cell lines;
- in the same vein, the expressed protein cannot readily be used for studies on induction of CTL responses, since the inoculation of the cell line into any host (other than that from which the parental line originated) would induce a massive allogeneic response which might obscure the specific response being studied.

The second technique, now the most widely adopted, is expression from a viral vector. The best system to date, and accordingly the most used, is based on vaccinia virus (VV), the prototype orthopox virus. The method of gene introduction and expression, developed in Bernard Moss' lab, has been extensively described; herein a summary of the technique is provided.

In principle, the gene of interest is cloned into a transfer plasmid, which is subsequently incorporated into the VV genome. This is summarized in *Figure 3*. A commonly used transfer plasmid is pSC_{11}. This has three notable components:

(a) An upstream (5') VV promoter (P 7.5) and a downstream (3') VV terminator flanking a single cloning site, for the restriction enzyme *Sma*I. The gene of interest (X in *Figure 4*) is cloned into this site in an appropriate orientation, and thus is flanked by appropriate VV transcriptional control signals. The gene should carry its own translational control signals (ATG and TGA/TAG/TAA).
(b) A marker gene encoding the enzyme β-galactosidase, under the control of a second VV promoter (P_{11}).
(c) These regions are flanked by sequences from the 5' end and 3' end of the VV thymidine kinase (TK) gene (hatched boxes in *Figure 4*).

The transfer vector is introduced (by calcium phosphate transfection) into cells infected with wild-type VV. Note that this virus contains an intact TK gene (represented as TK^+). At this point, control over the outcome evades the experimenter; one hopes for homologous recombination between the TK gene in the VV genome and the flanking TK sequences in the transfer plasmid. Should this occur, the resultant virus is a recombinant, which contains two new genes (β-galactosidase and the gene of interest) which can be transcribed by VV RNA polymerase but whose TK gene is bisected and therefore non-functional (represented as TK^-). The viruses harvested from this experiment comprise three groups:

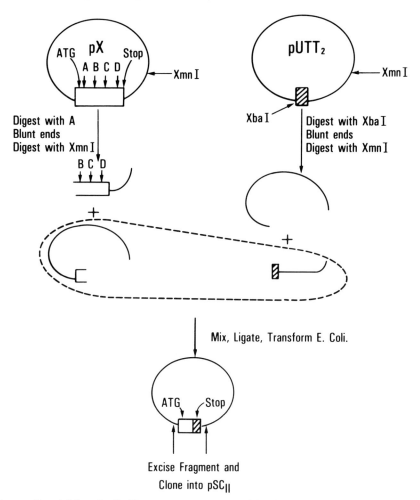

Figures 3 and 4 (overleaf). The strategy to use a vaccinia virus transfer plasmid. This allows transfer of the viral protein of interest to the target cell for CTL studies.

(a) input virus (VV_{wt}) which is TK^+,
(b) input virus (VV_{wt}) with spontaneous mutation in the TK gene and therefore TK^- (this occurs at a frequency of approximately 1 in 10^4), and
(c) recombinant VV which is TK^-.

This population is then subjected to two levels of selection simultaneously, to select firstly for TK^- virus [groups (b) and (c) above] and secondly for β-galactosidase expression [group (c)]. An experimental protocol for these procedures is given in *Protocol 7*.

Use in vitro to map CTL epitopes

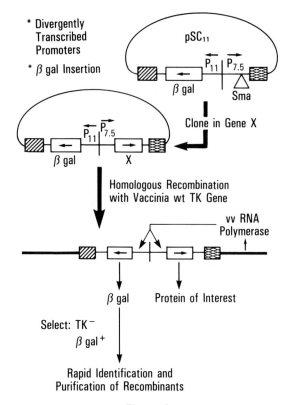

Figure 4.

Protocol 7. Expression of foreign genes in vaccinia virus

Materials required
- VV_{wt} at titre of $10^8 - 10^9$ p.f.u./ml
- 143TK⁻ cell line in culture
- HeLa cell line in culture
- modified Eagles medium (MEM) for cell culture
- bromodeoxy uridine (BUdR) dissolved at 25 μg/ml in MEM
- transfer plasmid (such as pSC_{11})
- gene of interest
- Xgal and dimethylformamide
- 2 × HBS
- 250 mM $CaCl_2$
- 100 mm tissue culture plate
- 1% sterile agarose
- 2 × 199 medium

Protocol 7. Continued

Methods

Cloning gene of interest into pSC_{11}

1. Prepare pSC_{11} DNA by routine alkaline lysis technique described elsewhere. Excellent guides to the routine molecular biology procedures required in steps 1–7 are available (e.g. Current Protocols in Molecular Biology: Published by Greene/Wiley).

2. Isolate gene of interest on as small a DNA fragment as possible. The gene must contain its own translational control signals (start and stop codons), and must be unspliced i.e. the parental gene must itself contain no splices, or else a cDNA copy of a spliced gene must be used (vaccinia virus genes contain no splices, and VV cannot process spliced genes since its life cycle is wholly cytoplasmic).

3. Linearize pSC_{11} with restriction enzyme *Sma*I.

4. Treat the linearized vector with calf intestinal phosphatase to remove the 5′ P groups, thus preventing religation of the vector in Step 5.

5. Mix vector and insert DNAs, ligate, and transform bacteria.

6. Screen colonies by miniprep analyses to identify one containing insert in desired orientation.

7. Prepare DNA of this plasmid, by same method as in step 1 above.

Introduction of plasmid into VV

1. Day −1. Plate 3×10^6 HeLa cells in 15 ml of MEM on to a 100-mm tissue culture plate. Grow overnight in a 5% CO_2 incubator, at 37 °C.

2. Day 0. Aspirate MEM from HeLa cells. Add 1 ml of MEM containing 10^6 p.f.u. of VV_{wt} (multiplicity of infection = ~0.2). Rock the plate gently for 45 min, ensuring no parts of the cell monolayer become dry.

3. While the plate is rocking, prepare DNA: add 10 μg of the plasmid DNA (from step 7 above) to 500 μl of 2 × HBS (560 mM NaCl, 100 mM Hepes, 3 mM Na_2HPO_4 pH 7.1) in a 4-ml snap cap tube and mix well. While holding the tube on a vigorous vortexer, add SLOWLY, DROP BY DROP, 500 μl of 250 mM $CaCl_2$. The end result is 1 ml of liquid containing an extremely fine granular precipitate.

4. After the plate has been rocked for 45 min, aspirate the inoculum. Add back 10 ml of fresh MEM, then drip the 1 ml of DNA precipitate into the MEM. Return to the incubator overnight to allow the DNA precipitate to settle on to the cell monolayer.

5. Day +1. Aspirate MEM from the plate. Wash gently ×2 with PBS. Add 10 ml of fresh MEM. Leave overnight in an incubator.

6. Day +2. By now, light microscopy will reveal marked cytopathic effect. Aspirate the MEM. Add 1 ml of fresh MEM, and use a cell scraper to remove

Protocol 7. *Continued*

cells from the plate. Transfer cells/MEM into a 2-ml cryovial. Freeze at −70 °C (dry ice) for 20 min, then place at 37 °C until fully thawed. Repeat this freeze–thaw cycle (which lyses cells without harming VV) twice more. Add 0.4 ml of 0.5% (5 g/litre) trypsin, vortex, keep at 37 °C for 15 min. Freeze at −70 °C. This treatment releases VV (which is largely cell-associated) and disaggregates the particles. Label the tube e.g. RM (Recombinant Mix), as it is a mixture of input and (hopefully) recombinant VV.

Identification of recombinant VV

1. Day −1. Plate out 12 100-mm plates, each with 3×10^6 143TK⁻ cells, in MEM containing 25 μg/ml BUdR.
2. Day 0. Prepare 4 ml of each of 10^{-2}, 10^{-3} and 10^{-4} dilutions of the RM (see step 6 above) in MEM/BUdR. Aspirate MEM from the 12 plates. Add 1 ml of 10^{-2} dilution to each of four plates; similarly for 10^{-3} and 10^{-4} dilutions. Rock for 45 min, aspirate the inoculum and add back 10 ml/plate of MEM/BUdR. Return to an incubator for 48 h. The BUdR prevents successful replication of TK⁺ virus, so only TK⁻ virus will replicate to form plaques. Visual/microscopic inspection after 48 h will reveal plaques scattered over the plates.
3. Day +2. Dissolve 45 mg of Xgal in 300 μl of dimethylformamide. Microwave 1% sterile agarose until melted. Combine 70 ml of agarose, 70 ml of 2 × 199 and the DMF/Xgal, mix gently but thoroughly. When 'hand-warm', aspirate MEM from the 12 plates, and add back 10 ml of the above mix to each plate (you will have a little excess left). Allow to solidify (20 min at room temperaure). Return to the incubator. After 5–6 h (sometimes you may have to wait overnight), recombinant virus plaques will appear as dark blue spots, due to expression of β-galactosidase which converts the colourless Xgal to a blue derivative.
4. Day +2/+3. Pick several (5–10) independent blue plaques. Lay the plate flat on the bench. Insert a sterile Pasteur pipette, with bulb, into the agarose, until it touches the plate. Aspirate the plug of agarose/cells/VV which should be deeply blue, and express into 1 ml of MEM. Take this through three cycles of freeze–thaw, and add 100 μl of trypsin; at 37 °C for 15 min then freeze at −70 °C.
5. The virus thus isolated may be 'contaminated' with TK⁻ non-recombinant particles. For this reason, three or four cycles of 'plaque purification' are advised. To plaque purify, prepare a 6-well plate with 143TK⁻ cells in MEM/BUdR. Plate 7×10^5 cells/well on day −1.
6. On day 0, carry out infection using 10^{-1}, 10^{-2} and 10^{-3} dilutions of the plaque-picked virus (in step 4 above) in duplicate on the 6-well plate. Thus, prepare 600 μl of each dilution and plate 300 μl into upper and lower adjacent wells. The top and bottom rows of the plate will be identical, each with a

Protocol 7. *Continued*

$10^{-1}/10^{-2}/10^{-3}$ well. Rock the plate for 45 min, aspirate, and replace with 3 ml of MEM/well.

7. After 48 h, aspirate the MEM from all wells. To each well of the top row, add 3 ml of agarose/199/Xgal mix as above. To each well of the bottom row add 1 ml of crystal violet stain (20% ethanol, 0.01% crystal violet, in water). When the agarose in the top row had solidified, draw off crystal violet from the bottom row, and wash these wells three times with 2 ml of PBS. Clear plaques will be seen against a dark background. Place the plate in an incubator until blue plaques appear in the top row. Count blue plaques (recombinant VV) and clear plaques (total VV). If "total" > "recombinant", pick an isolated blue plaque, and repeat steps 5, 6, and 7. When the counts are identical in the duplicate wells, then all VV are recombinants i.e. the virus is pure. At this point, large stocks of virus can be prepared for experimental analysis.

The advantages of this system are:
- it is now a fairly straightforward procedure,
- the reagent generated can infect a wide range (both within and between species) of cell lines thus allowing studies to be carried out on a wide variety of MHC backgrounds with a single recombinant, and
- similarly, studies of CTL induction can be undertaken, since the recombinant VV has also a broad *in vivo* host range.

A theoretical drawback is that one is analysing the antigenicity of the target protein against a background of VV infection, which is not its normal context of expression/presentation. To date most evidence points to this potential drawback being of little actual significance: all CTL epitopes identified using VV have been shown to be epitopes in the parental virus as well.

Many proteins from most virus families have been expressed in VV, with few failures. Their uses are manifold. In the context of this chapter, however, we wish to consider only their use in studies of CTL reactivity. The methods involved in induction of virus-specific CTL, their assay, and cloning, have been detailed in *Section 4.1*. The same principles apply to ^{51}Cr release assays using recombinant VV, but there are some differences in detail. *Protocol 8* provides a method for the preparation of VV infected target cells, and for production of anti-VV CTL. *Figure 5* demonstrates a typical set of results. Note in particular the presence of mandatory controls.

(a) There must be H2-mismatched virus infected targets, to ensure that lysis of the H2-matched virus infected targets by the effector cells is H2-restricted.

(b) There must be an uninfected cell control, and ideally cells infected with a different virus, to show that lysis is virus-specific. In practice, the control in (c) fulfills this latter function.

Use in vitro to map CTL epitopes

			\multicolumn{4}{c}{TARGET CELLS}		
			\multicolumn{4}{c}{H2^b}	\multicolumn{2}{c}{H2^d}	
			UN	LCMV	VV$_{NP}$	VV$_{SCII}$	UN	LCMV
H2^b	Anti LCMV	50:1	1	75	60	2	3	2
	Splenocytes	25:1	1	55	40	1	2	2
H2^d	Anti LCMV	50:1	2	3	3	2	4	68
	Splenocytes	25:1	1	1	3	3	2	59
H2^b	Anti VV	50:1	1	2	58	55		
	Splenocytes	25:1	1	3	49	50		

Figure 5. Typical experiments showing the virus-specific and H-2 (MHC) restriction using the vaccinia virus system. UN, uninfected; LCMV, lymphocytic choriomeningitis virus; VV, vaccinia virus; NP, LCMV nucleoprotein gene; SC$_{11}$, vaccinia vector with vaccinia insert only; anti-LCMV, LCMV-specific CTL; anti-VV, vaccinia virus-specific CTL. Ratio of splenic CTL to target cells was 50:1 or 25:1.

(c) As well as the recombinant VV expressing the gene of interest, another recombinant VV not expressing this gene must be used, to show that lysis is gene-specific. For this purpose we routinely use a recombinant VVSC$_{11}$ in which the plasmid pSC$_{11}$ itself (with no insert gene) was recombined into VV.

(d) A conclusion of gene-specific killing (in the case of *Figure 5*, that anti-LCMV splenocytes see the NP molecule) rests on the fact that VVSC$_{11}$ infected cells are not killed while VV$_{NP}$ infected cells are killed by the anti-LCMV splenocytes. Thus it is important to confirm that the VVSC$_{11}$ and VV$_{NP}$ cells are equivalently VV infected; this purpose is fulfilled by having effector cells specific for VV, as shown in *Figure 5*.

Protocol 8. *In vitro* cytotoxicity assays using VV

The basic outline of the *in vitro* cytotoxicity assay is provided in Section 4.1.

Preparation of VV infected target cells

1. Seed 2×10^6 H2-matched ($\pm$ H2-mismatched) cells on to a T75 flask, with 20 ml of MEM, and grow overnight.
2. The evening prior to assay, infect the cells with the appropriate recombinant VV, at multiplicity = 3. Aspirate the MEM and replace it with 1 ml of MEM containing approximately 10^7 p.f.u. VV. After gentle rocking for 45 min, replace the inoculum with 20 ml of MEM, and return the T75 to the incubator overnight.
3. The following morning, wash the cells twice with PBS, and remove them from the flask using trypsin. Wash cells once with MEM, and count them. Label the

Protocol 8. *Continued*

desired number of cells with ^{51}Cr as described earlier, wash them, and use them as targets.

Preparation of anti-VV effector cells

1. Inject two mice intraperitoneally with 2×10^7 p.f.u. of $VVSC_{11}$.
2. 6–7 days later sacrifice the animals and remove their spleens. Harvest effector cells as described for anti-LCMV effectors in Section 4.1.

Using these techniques it has been convincingly shown that most viral proteins can act as CTL target antigens. It is, however, important to realize that the target antigen will vary enormously depending on the host MHC molecules. For example, C57BL/6 mice ($H2^b$) mount brisk CTL responses to both GP and NP components of lymphocytic choriomeningitis virus: in contrast, BALB/c mice ($H2^d$) mount a measurable response only to the NP moiety.

9. Mapping viral epitopes seen by CTL

Having identified the viral protein seen by CTL, the next step is to identify more precisely the target residues involved in CTL recognition. Two general methods have been employed—first, the expression of fragmented proteins using VV, and secondly the administration to uninfected cells of short synthetic peptides.

(a) The generation of recombinant VV has been described. In principle the expression of protein fragments differs only in extra cloning steps to manufacture a fragment which contains its own translational control signals. Several techniques may be used, and one used successfully in our laboratory is diagrammed in *Figure 3*. The technique allows the construction of a family of C-terminally truncated proteins. The method for the cloning procedure is given in *Protocol 9* and requires a plasmid $pUTT_2$ (J. L. Whitton, unpublished) which encodes a universal translation terminator (UTT) sequence immediately downstream of a unique *Xba*I site. The UTT encodes a translational termination codon in each of the three reading frames, and thus ensures that any open reading frame (ORF) placed upstream of this sequence will not encode a lengthy fusion protein. This plasmid is available from J. L. Whitton. The recombinant VV encoding the truncated proteins can then be used to provide target antigens for CTL clones, thereby analysing the epitope specificity of these cells. VV containing the epitope will be recognized by the CTL clone, while truncations in which the epitope is removed will remain immune to CTL lysis. In this way a region of protein (its size determined by the distance between adjacent truncation end points) is defined as critical for CTL recognition. At this point, the further characterization of the epitope is carried out using the second technique (below).

Protocol 9. Generation of 3' serial deletions in an ORF

Materials required
- Plasmid pUTT$_2$ (see text and *Figure 3*).
- Plasmid encoding gene of interest (pX in *Figure 3*).
- Restriction enzymes, DNA ligase, etc. for cloning.

Method

1. Map plasmid pX for restriction sites, and select for use ideally those enzymes cutting uniquely, and within the open reading frame (ORF) of the gene of interest. Such enzyme sites on pX are labelled A, B, C, and D (*Figure 3*). The enzyme *Xmn*I must not cut within the gene.
2. Digest pX with one of the selected enzymes (e.g. A in *Figure 3*), make the terminus blunt (if necessary) and recut the plasmid with *Xmn*I to generate a large fragment containing the 5' ORF sequences including the ATG. This should then be gel-purified.
3. Digest pUTT$_2$ with *Xba*I, blunt the termini, and recleave the plasmid with *Xmn*I. Gel-purify the small fragment containing the UTT.
4. Mix, ligate the two fragments purified in 2 and 3, and transform into *E. coli*. In theory, the only possible viable plasmid will be one in which the 5' ORF (with ATG) will be apposed to the UTT; this plasmid thus contains an ORF encoding an N-terminal fragment of the parental protein.

(b) Exact identification, at the single amino acid level, of a CTL epitope requires the use of synthetic peptides. While it would be possible to identify epitopes by using synthetic peptides alone (rather than first identifying a restricted area using the truncations described above) it is not advisable because

(a) it would require the synthesis of overlapping peptides across a whole protein, which is very costly, and

(b) even with overlapping peptides there is a real risk that no peptide would contain the complete epitope(s), so it may be missed.

Thus in practice the identification of the epitope is best done by a judicious combination of recombinant DNA and synthetic peptide technology.

In the first instance, synthetic peptides should be made spanning the 'critical' sequence. Each peptide should be 12–16 residues in length, and peptides should overlap by five residues. Peptides are incubated with uninfected H2-matched and H2-mismatched target cells, which are then incubated with the CTL clone(s). Details are given in *Protocol 10*. In our experience there is little point in trying to predict from the primary sequence those regions likely to be CTL epitopes. Furthermore, one should not become discouraged about whether a given peptide sequence will dissolve in MEM: even extremely hydrophobic peptides can be successfully used to sensitize target cells.

Protocol 10. Use of peptides to sensitize for CTL lysis

Carry out an *in vitro* cytotoxicity assay as described, with the following additions.

1. Incubate H2-matched and H2-mismatched uninfected cells with 40 µg of peptide/well. Dissolve each peptide in MEM at a final concentration of 800 µg/ml and add 50 µl to each appropriate well. Add 2×10^4 ^{51}Cr-labelled target cells to each well in 50 µl, and add effector cells (CTL clones) in 100 µl. Thus the final concentration of peptide is 200 µg/ml. (For a 10-residue peptide, this is ~ 200 µM). The assay thereafter is as described.

2. When a positive peptide is identified by the above method, generate a dose-response curve by using peptide at 10-fold dilutions to identify the peptide concentration giving half-maximal lysis (often < 1 µM).

When a positive peptide is identified, it is then possible to continue the analysis in two ways. Firstly, nested sets of peptides can be made with N-terminal and C-terminal truncations of the original, to identify the minimal residues needed for recognition. Secondly, amino acid substitutions can be made to dissect the importance of the individual residues. These two procedures identify the fine specificities of CTL clones, and often will allow one to distinguish different specificities amongst clones which see the same nominal epitopes.

4D. USE *IN VIVO*
LINDA S. KLAVINSKIS and MICHAEL B. A. OLDSTONE

10. Introduction

The development of techniques for *in vitro* cloning and long-term culture of antigen-specific T cell clones has made it feasible to obtain large numbers of homogeneous cloned T cells and study their specific functions *in vivo*. These homogeneous populations have been used to ask questions concerning the specific contributions of CTL in the control of viral infections (1–5). The information obtained to date has indicated that for several viruses such as influenza, lymphocytic choriomeningitis, murine cytomegalo and respiratory syncytial, CTL alone are sufficient to control the viral infection (1–5). This promises to be of enormous practical value for the potential use of cloned CTL in immunotheroapy against chronic viral disease and in considering the incorporation of T cell epitopes in the future development of synthetic vaccines. CTL clones have also been of value in demonstrating the adverse role of CTL (under certain circumstances) in the induction of immunopathological disease (5, 6). In addition, CTL clones have also been of value as unique probes for defining CTL trafficking and target cell homing *in vivo*, as well as illustrating the molecular basis of CTL-mediated cytolysis *in vivo* (3, 7). For example, in mice dually infected with two virus subtypes, it has been shown that the subtype specific CTL clone only reduces virus titres of the recognized virus subtype, thus indicating that the anti-viral activity of CTL *in vivo* is highly specific and implying that CTL express their effector activity *in vivo* by direct cytolysis of infected cells (3). Methods for preparing murine T cells clones for adoptive transfer *in vivo* will be discussed below.

11. Preparation of CTL clones for *in vivo* studies

11.1 General considerations

The major considerations for obtaining reproducible results with *in vitro* propagated CTL clones (or T cell lines) for adoptive transfer experiments are:

(a) *Viability and activation state* in vitro. Unless the cloned CTL are morphologically viable and proliferating *in vitro*, such cells are unlikely to survive *in vivo*.

(b) *IL-2 dependence.* It has been suggested that clones highly dependent upon exogenous IL-2 may fail to proliferate *in vivo*, or at best give irreproducible results. The addition of 150 IU of recombinant murine IL-2 (Genzyme, Boston, MA) or 100 ng of recombinant human IL-2 (Sandoz, Vienna, Austria; 8) to the cells for transfer has in some cases prolonged the survival of cloned T cells *in vivo* (8, 9). At best, the addition of recombinant IL-2 tends to increase the reproducibility of clonal T cell transfer experiments.

(c) *Duration of* in vitro *culture.* Long-term culture of T cell lines and cloned CTL (over several months) has been associated with morphological changes in size and granularity (10) and with the loss of certain cell surface receptors essential for normal lymphocyte migration (11).

(d) *Removal of non-viable cells.* Damaged and dead cells tend to be very sticky probably because of the adherent DNA they release which causes aggregates of cell debris and hence problems upon injection *in vivo*.

12. Removal of dead cells

There are a number of simple procedures for removing irradiated feeder cells, antigen presenting cells (APCs), and non-viable CTL (see *Protocol 11*). In our experience, method (C) is simple, with minimal loss of cloned T cells.

Protocol 11. Removal of dead cells

Centrifugation through Ficoll-Hypaque.

1. Harvest both the adherent and non-adherent (feeder/APC's) by trituration.
2. Centrifuge at 400 g for 5 min
3. Resuspend the cell pellet to $5 \times 10^6 – 2 \times 10^7$ cells/ml in RPMI-1640.

Protocol 11. *Continued*

4. Layer 2–6 ml of the cell suspension to 4 ml of Ficoll-Hypaque (density = 1.09).
5. Centrifuge at 2000 g for 20 min

The interface can be recovered with high viability but with some loss of live cells. The technique also requires that the cells be subsequently washed at least three times in RPMI-1640 supplemented with 5% FCS (250 g for 10 min) to remove the Ficoll.

Centrifugation through fetal calf serum.
1. Carefully layer 3 ml of FCS under 1–2 ml of the cell suspension (at 5×10^6–2×10^7 cells/ml) in a 10-ml centrifuge tube.
2. Centrifuge at 350 g for 10 min
3. Remove the supernatant, and resuspend the cell pellet.

Differential removal of feeder cells from cloned CTL.
This method relies upon the semi-adherent characteristics of established CTL clones *in vitro*. The advantage of this procedure is that it is quick and simple; removes both dead cells and feeder cells from the cultures, and requires only one or two centrifugal washes, thus minimizing loss of CTL associated with density gradients and rinsing procedures.

1. Gently aspirate the medium from the cultures.
2. Wash the adherent CTL (twice) with plain RPMI-1640 and agitate the plate/flask gently (by tapping) to dislodge any residual non-adherent cells.
3. Aspirate one final time and then add tissue culture grade trypsin-EDTA (0.05% w/v trypsin [1:250] and 0.02% w/v EDTA, Flow Laboratories, McLean, Va.). Generally, 3–4 drops from a pasteur pipette to a well of a 24-well plate, or 1 ml to a 9-cm petri dish will suffice. Gently agitate the plate (1–2 min) and check for dissociation of cloned CTL with an inverted microscope.
4. Immediately neutralize the trypsin solution with RPMI-1640 supplemented with 10% FCS (0.5 ml per well of a 24-well plate, or 5 ml to a 9-cm petri dish). Sediment the pooled suspension at 400 g for 5 min, resuspend, and rewash once in RPMI-1640 containing 5% FCS.
5. Count the number of viable cells (with trypan blue/acridine orange/ethidium bromide) and adjust to the required cell concentration per 0.4 ml volume for subsequent injection.

These simple procedures should ensure that the cloned CTL suspension does not contain cell debris or aggregates of dead cells. It is essential that all such aggregates are removed, otherwise intravenous infusion of suspensions containing aggregates of cells will induce respiratory distress, stroke, cerebral infarction and death of the recipient. It is often useful to filter the cell suspension (just prior to *in vivo* inoculation) through a layer of prewetted fine nylon gauze, cotton bandage or a small plug of teased adsorbent cotton wool in a filter funnel over a 10-ml centrifuge tube. Spin the suspension at 400 *g* for 5 min and resuspend the cell pellet in RPMI-1640 containing 5% FCS.

13. In vivo procedures: immunosuppression

When examining the role/activity of adoptively transferred CTL *in vivo*, e.g. the contribution of cloned CTL in the control of virus infection, it is essential to immunosuppress the recipient, to dissociate the contribution of the recipients immune response. This can be successfully accomplished using either irradiation or immunosuppressive drugs.

(a) *Irradiation*. For light irradiation, 450–600 rads from a ^{137}Co-source is generally sufficient and 800 rads for sublethal irradiation. Doses in excess of 900 rads are generally lethal! Strains of mice do vary in their susceptibility to ionizing radiation, e.g. in order of susceptibility: BALB/c > SWR/J > C57BL/6.

(b) *Immunosuppressive drugs*. Cyclophosphamide (Cytoxan, Mead, Johnson, Evansville, IN) is commonly used to suppress the immune response of recipients prior to cell transfer. Both the dose and time of administration of this drug are critical for prevention of inflammation. Treatment with high doses (300 mg/kg i.p.) is effective in short term experiments where the immediate natural killer cell, or CTL response of the recipient, is suppressed. However, this dose of drug does allow the later development of a CTL response originating from the recipient, if prior inoculated with virus. In circumstances where it is necessary to delay the transfer of CTL for several days following virus infection of the recipient, e.g. to study induction of immunopathology mediated by the transferred CTL (11), it has been found that a single i.p. dose of 175 mg/kg given 4–5 days after infection of the recipient is sufficient to prevent the development of a recipient-derived CTL response within 4 days of cell transfer.

14. In vivo inoculation

To examine systemic effects of cloned CTL *in vivo* or study their migration, it is essential to inoculate via the intravenous (i.v.) route; either via the orbital plexus or tail vein. It is generally helpful to dilate the tail veins under a heating lamp or in lukewarm water, and *slowly* inject the required cell concetration in a volume of 0.3–0.4 ml, using 27 gauge needles. For studying local effects of CTL, e.g. in the central nervous system or the induction of delayed type hypersensitivity in the footpads of mice, a maximal volume of 0.3 ml injected intracerebrally or into the footpad is recommended.

15. Migration studies *in vivo*

Radiolabelling cloned CTL

Ideally radioisotopes used to measure the migration and activity of adoptively transferred CTL should be non-toxic, easily measured, remain bound to cells for the duration of the experiment and not be re-utilized upon cell death. Practically, none of the available radioisotopes confirm to all these specifications, however ^{51}Cr and ^{111}In, at low doses of radioactivity are relatively non-toxic and widely used.

Protocol 12. Migration studies *in vivo* with radiolabelled cloned CTL

1. Harvest and prepare viable cloned CTL (Section 11) and adjust to 1×10^7 cells/ml in complete medium.

2. Incubate for 1 h at 37 °C with 25 μCi/ml of ^{51}Cr (sodium chromate) in aqueous solution (sp. act. 1 mCi/ml, Amersham, Inc.) This should result in cells labelled to approximately 50–100 c.p.m./10^4 cells, minimizing radiotoxicity.

3. Wash the cells by centrifugation at 400 g for 5 min. Resuspend the cell pellet in complete medium and recentrifuge until the radioactivity in the supernatant is <2% of that in the cell pellet. This usually requires 4–5 washes.

4. Resuspend the cells to $1 \times 10^7/0.4$ ml and inoculate i.v. (Section 5). Retain an aliquot of the injected cells as a reference to calculate the *in vivo* localization of the injected radioactivity.

5. To measure cell distribution, sacrifice the recipients at various time intervals, remove organs and quantitate the radioactivity using a gamma counter.

6. Calculate the percentage localization of the injected radioactivity with reference to the retained aliquot of labelled cells.

---

16. Measurement of the ability of cloned CTL, as compared with freshly isolated lymphocytes, to migrate to lymphoid tissue

T cell clones maintained in culture with IL-2 and restimulated with antigen tend to lose their ability to bind to high endothelial venules (9) and thus may be markedly deficient in their ability to home to peripheral lymphoid tissue, and may thus accumulate in the liver and lungs. One way of assessing the migratory activity of cloned CTL relative to freshly isolated lymphocytes, e.g. splenocytes, involves monitoring the migration of ^{51}Cr-labelled CTL with ^{111}In-labelled normal spleen cells.

Protocol 13. Measurement of migratory activity of cloned CTL

1. Prepare a suspension of freshly isolated normal spleen cells and adjust to 1×10^7 cells/ml.
2. Label spleen cells for 20 min at room temperature with 20 µCi/ml ^{111}In (sp. act. >10 mCi/µg, Amersham, Inc.).
3. Wash the cells four times at 400 g for 5 min.
4. Adjust ^{111}In-labelled cells to 1.5 times the c.p.m. of ^{51}Cr-labelled CTL.
5. Adjust the volume of ^{51}Cr-labelled cells with ^{111}In-labelled cells to a final volume of 0.3–0.4 ml and inject i.v. (retain an aliquot of ^{51}Cr and ^{111}In-labelled cells).
6. Sacrifice the recipients at the required time points and quantitate the two radioisotopes in a two-channel counter.
7. The relative localization ratio (RLR) for each organ is a measure of the relative localization of cloned CTL compared to the infused freshly isolated splenic lymphocytes, after adjustment for radioactive decay.

References

1. Byrne, J. A. and Oldstone, M. B. A. (1984). *J. Virol.*, **51**, 682.
2. Lin, Y. L. and Askonas, B. A. (1981). *J. Exp. Med.*, **154**, 225.
3. Lukacher, A. E., Braciale, V. L., and Braciale, T. J. (1984). *J. Exp. Med.*, **160**, 814.
4. Reddehasse, M. J., Mutter, W., Munch, K., Buring, H-J., and Koszinowski, U. H. (1987). *J. Virol.*, **61**, 3102.
5. Cannon, M. J., Openshaw, P. M., and Askonas, B. A. (1988). *J. Exp. Med.*, **168**, 116.
6. Klavinskis, L. S., Tishon, A., and Oldstone, M. B. A. (1989). *J. Immunol.*, **143**, 2013.
7. Young, L. Y., Klavinskis, L. S., Oldstone, M. B. A., and Young, J. D-E. (1989). *J. Exp. Med.*, **169**, 2159.
8. Taylor, P. M. and Askonas, B. A. (1987). *Immunology*, **58**, 417.
9. Dailey, M. O., Fathman, C. G., Butcher, E. C., Pillemer, E., and Weissman, I. (1982). *J. Immunol.*, **128**, 2134.
10. Taylor, P. M. and Askonas, B. A. (1983). *Eur. J. Immunol.*, **13**, 707.
11. Baenziger, J., Hengartner, H., Zinkernagel, R. M., and Cole, G. A. (1986). *Eur. J. Immunol.*, **16**, 387.

5

Methods for studying mouse natural killer cells

RAYMOND M. WELSH

1. Introduction

Natural killer (NK) cells are bone marrow-derived, thymus-independent, large granular lymphocytes (LGL) which, without specificity or antigenic memory, lyse on contact a variety of target cells (1, 2). The prototypic NK cells lack T cell receptor or CD3 mRNA or proteins (3). NK cells lyse a variety of tumour cells and certain normal cells such as thymocytes. Infection of target cells with any of a variety of viruses contributes to enhanced NK cell-mediated lysis of those targets. The cytotoxic capacity of NK cells can be greatly augmented by interferon (α, β, or γ) and by interleukin-2 (IL-2), resulting in enhanced kinetics of lysis of highly sensitive targets and a greatly expanded target range encompassing most cultured cell lines. Interferon and IL-2 also contribute to the proliferation of NK cells. Virus infection, as a result of stimulating very high levels of IFN (α and β), causes marked activation and proliferation of NK cells *in vivo*. Studies in the mouse have characteristically shown a peak in NK cell activity and number early in infection, followed by a peak in cytotoxic T cells later in infection (2). In keeping with the general orientation of this book, most of the following discussion will focus on NK cells in the mouse, but many of the techniques and general principles discussed can also be applied to studies in man and other animals.

NK cells bind to target cells via ill defined receptors in a calcium cation-dependent process (1). Following the binding event there is a triggering event (possibly due to phospholipase and transmethylase activities) which causes a calcium flux into the NK cell cytoplasm. This elicits a reorientation of the NK cell such that its cytoplasm is proximal and nucleus distal to the target cell. Granules flow from the golgi and align themselves at the membrane adjacent to the target cells. The granules, which contain a cytotoxic molecule known as perforin or cytolysin as well as serine esterase, are secreted (4). The perforin/cytolysin, which is structurally related to the ninth component of the complement system, oligomerizes into a ring-like structure on the target cell membrane, possibly resulting in the osmotic disruption of the target. Other factors may also be involved in cytotoxicity, such as the lymphokines tumour necrosis factor (TNF) and lymphotoxin and one less-defined product called NK cell cytotoxic factor (NKCF). Inconsistent with the concept that osmotic disruption is the main cause of cell death is the finding that prior to death the target cell DNA is degraded, frequently correlating with the generation of nucleosome-sized fragments. Of interest is that the treatment of target cells with IFN tends to render them resistant to NK cell-mediated lysis at a post binding step.

Considerable evidence has indicated that NK cells may provide natural resistance against virus infections and may even participate in classical immune mechanisms by using their Fc receptors and anti-viral antibody produced by B cells to mediate virus-specific antibody-dependent cell-mediated cytotoxicity (ADCC) (2).

2. Histochemical staining

NK cells displaying the classic LGL morphology can be visualized by Wright's Giemsa staining. They can be seen in blood smears, but best results are seen with cells pelleted on to glass slides in a cytocentrifuge. Slides are inserted into slide racks and immersed into a staining jar of freshly filtered Wright's stain for 4 min, raising and lowering the slides several times to cause a gentle mixing. These are then transferred to a staining jar containing Sorenson's buffer for 1 min with gentle mixing and then transferred into a vessel containing freshly filtered Giemsa stain for 4 min. These slides are then washed once more in Sorenson's buffer and air dried.

2.1 Reagents

Wright's stain. Wright's stain (2.5 g/litre in methanol, pH 6.7, Sigma Chemical Co., St. Louis, MO) is used undiluted, but must be filtered before each use. The Wright's stain can be reused up to five times if it is properly removed from the staining jar and stored in an airtight container to prevent methanol evaporation.

Sorenson's buffer. A stock solution is prepared by dissolving 9.08 g of KH_2PO_4 and 2.97 g of anhydrous Na_2HPO_4 (or 5.98 g of $Na_2HPO_4.12H_2O$) into 800 ml of distilled water. The pH is adjusted with 10 N NaOH to 7.2 and the final volume is adjusted to 1250 ml. This is stored at 4 °C. The working buffer solution is made by mixing 25 ml of stock Sorenson's with 475 ml of distilled water.

Giemsa stain. Diluted Giemsa is prepared by filtering 12.5 ml of stock solution (7.415 g/litre in methanol, Accra Lab, Bridgeport, NJ) into 10 ml of concentrated Sorenson's buffer plus 230 ml of distilled water. The stain is mixed well, used fresh the same day (within 8 h), and discarded after use.

Filter paper. Use Fisherbrand 12.5 cm, P8 coarse porosity, #09-795E.

2.2 Interpretation

NK LGL characteristically are large lymphocytes with a reniform or kidney bean-shaped nucleus and a relatively large cytoplasm usually not exceeding the nucleus in size. Within the cytoplasm are red granules. LGL exhibiting blast cell morphology have enlarged cell size, a basophilic (dark blue-staining) cytoplasm and prominent nucleoli; some contain mitotic figures.

3. Cytotoxicity assays

3.1. Chromium-release assay

The most commonly used assay for NK cells and for CTL is the ^{51}Cr-release microcytotoxicity assay. Essentially the same assay can be used with ^{111}In, if this very short half-lived isotope is available to the investigator. Pellets of target cells are mixed with $Na_2^{51}CrO_4$ (in aqueous solution at 1 mCi/ml) at a ratio of 50–100 uCi/10^6 cells and incubated for 1 h at 37 °C. The pellets are centrifuge washed 2–3 times in 5 ml of tissue culture media [such as RPMI or Eagle's minimal essential medium (MEM) or other suitable buffer] and resuspended at 10^5 cells/ml in assay medium. Assay medium is the tissue culture medium supplemented with 5–10% heat-inactivated (56 °C, 30 min) FCS. The inclusion of 0.01 M Hepes buffer (N-2-hydroxyethylpiperzine-N'-2-ethane sulphonic acid; Sigma Chemical Co.) helps to maintain the pH. The target cells are dispensed in 100-μl aliquots into wells in microtitre tissue culture plates (300 μl volume wells). Either flat, round, or conical bottom wells may be used. The flat-bottomed wells tend to give more linear cytotoxicity responses, but the round and conical bottom wells tend to reduce the spontaneous lysis of the targets. Effector cells are prepared in assay medium and plated in 100-μl aliquots in quadruplicate on to the target cells. Usually at least three different dilutions of effector cells should be used to generate several effector to target ratios, such as 100, 50, and 25:1 or 100, 33, and 11:1 etc. No more than 2×10^6 effector cells (200:1) should be used per well, as the crowding of cells will tend to reduce cytotoxic effects at higher effector to target ratios. Obviously, the more cytotoxic (or purified) the effector cell population, the lower the effector to target ratio required. For each set of target cells controls must be set up to determine the spontaneous release of radiolabel and the maximum possible release of radiolabel. For this, one set of target cell wells should be treated with assay medium only, whereas the other set should be treated with 100 μl of 1% Nonidet P-40 in water shortly before harvest. Assays should run from 4 to 16 h. In general, the convenient overnight assays yield higher cytotoxicity determinations, but also higher spontaneous release of isotope and complications in interpretation of data due to leukocyte interactions and cytokine secretions during the assay period. Cytotoxicity values are lower in the 4-h assays, but they are cleaner and more interpretable and are therefore preferred.

For harvest, the microtitre plates are centrifuged at low speed to pellet the cells. Several centrifuge companies provide adapters to spin microtitre plates. If none are available, the assay can still be harvested if the investigator avoids shaking the plates. Automatic harvesting of microtitre plates can be done by instruments available from several companies. Otherwise, the investigator simply pipettes off

100 μl from the supernatant of each well and counts the sample in a gamma counter.

Analysis of data

1. *Calculation of lysis.* The percentage specific lysis of the target cells is determined by the formula

$$\% \text{ lysis} = \frac{\text{average c.p.m. Test} - \text{average c.p.m. MEM}}{\text{average c.p.m. NP-40} - \text{average c.p.m. MEM}} \times 100$$

2. *Calculation of lytic units (standard method).* The lytic unit calculation of Cerrotini (5) is commonly used as a way to standardize cytotoxicity data and to avoid the bias that can result by selection of single effector to target ratio to give as a cytotoxicity determination. When several effector to target ratios are used, a cytotoxicity curve can be generated with the y axis representing the percentage lysis and the x axis the effector cell number. When plotted as a semi-log plot, cytotoxicity curves are usually linear in the 20–30% cytotoxicity range. A given percentage within this range is selected (e.g. 20%), and this point on the graph is extrapolated down to the x axis, and the effector cell number required to lyse those effector cells is designated as one lytic unit. Dividing the total effector cell number by this number gives one the number of lytic units per cell population. The reader is also referred to statistical programs that can be used to calculate lytic units from cytotoxicity cuves (6).

3. *Calculation of lytic units (alternative method).* The standard lytic unit calculation offers some disadvantages in that it requires that the effector cell population mediates a rather high level of cytotoxicity in order to get at least a 20% level on a cytotoxicity curve. Sometimes it is important to compare relative levels of cytotoxicity in cell populations mediating low levels of lysis. A short cut method to give a rough estimate of comparative lytic units in such preparations uses the equation:

1 lytic unit = fraction of target cells killed/effector:target ratio.

This number multiplied by the total number of leukocytes in a preparation will equal the total lytic units in the sample. The same effector to target ratio should be used in all comparative samples (giving the advantage of not having variables caused by crowding effetcs with different numbers of effector cells). As much as possible, an effector to target ratio on the linear portion of the cytotoxicity curves should be used.

3.2 Choice of target cells

Virtually any kind of target cell can be used in microcytoxicity assays, but their sensitivities to NK cells will vary greatly. For the mouse and rat, the YAC-1 lymphoma is the prototype highly NK-sensitive target, although RL♂1 cells have nearly comparable sensitivity. The K562 cell line is most commonly used in human studies. Natural primary targets for NK cells include a subpopulation of large, cortisone-sensitive thymocytes (7). NK cells activated by IFN or IL-2 lyse a wide range of targets, any of which may be used in the cytotoxicity assays.

3.3 Single cell cytotoxicity assay

The single cytotoxicity assay, originally developed by Grimm and Bonavida (8) allows one to visualize cell death and to determine the proportion of effector cell–target cell binding events which lead to lysis of the target. Here target cells are identified by their large size, and dead cells are identified by uptake of the dye trypan blue.

3.4 Selection of target cell

As with the Cr-release assay, any of a variety of target cells can be used, dependent on the experiment, but for simple quantitation of NK cell activity, either K562 or YAC-1 cells are preferred. K562 cells offer the advantage of large cell size, allowing for easy target cell identification, and good stability within the assay. However, K562 cells are only moderately sensitive to endogenous NK cell activity, rendering them useful mostly in studies with activated NK cells. YAC-1 cells offer the advantage of high sensitivity to endogenous NK cells, but they are smaller cells not always easily distinguished from large blast-size NK cells, and they have to be carefully grown in log phase to withstand the rigours of the assay.

Protocol 1. Assay procedure

1. In a conical centrifuge tube, mix 6×10^5 target cells with effector cells at a ratio of 1–10 effectors/target in a total of 2 ml of complete tissue culture medium, such as RPMI.

2. Centrifuge at low speed (1000 r.p.m.) for 5 min to form a pellet. Draw off most of the supernatant, and allow the undisrupted pellet to incubate at room temperature for 30 min. This allows for efficient binding but not lysis of the target cell (incubation at 4 °C moderately inhibits binding, and incubation at 37 °C results in effector cells completing the lytic process).

3. Make 1% agarose (1:1 type I and type VII; Sigma Chemical Co.) solution in water. Melt this in a microwave oven and allow to equilibrate at 39 °C.

4. Dilute the agarose 1:1 with $2 \times$ RPMI assay medium, and resuspend the pellet in about 0.5 ml of the agarose–RPMI solution.

5. Layer a thin layer of agarose on to a microscope slide, and bake in a 75 °C oven for 2–3 h. This allows for better adhesion of the agarose layer in the next step.

6. Pipette 2 ml of agarose–$2 \times$ RPMI on to the slide to form a layer, and allow it to harden (precoating).

7. Pipette the resuspended pellet on to the slide (two slides/pellet) and allow it to harden.

Protocol 1. *Continued*

8. Submerge the slide in a vessel containing RPMI assay medium and incubate for 5 h in a 37 °C CO_2 incubator.
9. To visualize dead cells, remove the slides from the RPMI and submerge in 0.1% trypan blue in PBS for 5 min in a staining dish with mixing.
10. Transfer these to another staining dish and given three 5-min washes with PBS. Then submerge them in PBS containing 0.1% formaldehyde and incubate at room temperature overnight.
11. Allow the slides to dry on the bench top and examine them under a light microscope. Effector cell-target cell complexes are easily seen, and the dead target cells can be identified by their assimilation of the blue dye.

3.5 Modification of the single cell assay to quantitate lysis mediated by blast cells

The single cell assay can be adapted to measure lysis mediated by thymidine-incorporating blast NK cells (9). Effector cells are first pulsed with ^{3}H[thymidine] (25 μCi/10^7 effector cells/ml) for 1 h. This labels the DNA in dividing cells. The rest of the assay proceeds as above, but after fixing in formaldehyde the slides are briefly dipped in Kodak emulsion type NTB2 at 43 °C. The slides are dried and kept in the dark at room temperature for 2 days, then developed by submerging in Kodak D-19 diluted 1:1 with water for 4 min. After a 10-sec water wash, the slides are fixed for 5 min in Kodak fixer, washed in water for 5 min, and then examined under the light microscope. [^{3}H]thymidine incorporating cells are identified by the presence of black grains over the cell body.

4. Isolation and preparation of effector cells

A wide variety of techniques are available for the isolation of effector cells from different tissues. In general, single cell suspensions are made of leukocyte preparations, and erythrocytes are removed by a brief (1–2 min) suspension of cells in 0.84% NH_4Cl in water, followed by low speed centrifugation to pellet the cells. The resulting pellet should be immediately resuspended in a suitable buffered wash medium (such as RPMI, MEM, or Hank's buffer without serum), pelleted, and resuspended in assay medium. Mouse NK cells are heat labile, and it is wise to do all manipulations at 4 °C, except for the erythrocyte lysis procedure, which should be done at room temperature. Below are suggestions to isolate effector cells from different organs.

Spleen
Spleens are excised and minced either with forceps or between the frosted edges of two microscope slides. The cell suspension is filtered through a 202-μm pore size

nylon mesh (Tetko #HC3.202) to eliminate clumps and pieces of the spleen capsule. The cells are pelleted and exposed to the 0.84% NH_4Cl solution to lyse erythrocytes. Sometimes clumps of DNA appear after this treatment. If desired, these can be removed by filtration through the nylon mesh or by treatment with DNase at room temperature for 20 min.

Peritoneal cavity

Leukocytes are isolated from the peritoneal cavity by peritoneal lavage. Here, mice are injected intraperitoneally with 4–6 ml of media containing serum; the inclusion of herapin at 20 U/ml is sometimes used to inhibit clotting. The abdomen is palpated to suspend the leukocytes, and the fluid is drawn out with the syringe. Sometimes debris clogs the syringe needle. An alternative way to harvest the cells is to make an incision in the lower abdomen and pull back the skin and fur, leaving the peritoneal membrane exposed. The peritoneal cavity can then be entered with a Pasteur pipette, and the peritoneal leukocytes can then be easily collected.

Peripheral blood

Peripheral blood leukocytes (PBL) normally have too high a proportion of erythrocytes to use NH_4Cl as an initial purification step. PBL are usually isolated by sedimentation on gradients of percoll, ficoll, or other high density solution. There are several commercially available reagents specifically designed for PBL isolation.

Lymph nodes

Extraction of leukocytes from lymph nodes is similar to that of the spleen.

Bone marrow

The femur is removed, and its ends are cut off to expose the hollow bone barrow. A syringe containing RPMI medium is placed at one end, and the medium is injected into the bone marrow, causing cells to come out the other side.

Liver

The liver is not a normal lymphoid organ, but high levels of NK cells can accumulate within its parenchyma during viral infection (10). For best results the liver must be perfused. This will remove leukocytes from the blood vessels feeding the liver and will allow for proper enzymatic digestion of the liver, yielding the parenchymal liver leukocytes. The mouse should first be anaesthetized by an intraperitoneal injection of 1–2 mg of sodium pentobarbital. The mouse should be secured face-up on a dissecting board, and the abdominal cavity opened by a ventral midline incision. Lift out the intestines to expose the portal vein. Place a silk suture around the portal vein about 0.5 cm from the liver. Using microscissors make a small hole in the side of the portal vein about 1 cm from the liver. The ligature should be between the hole and the liver. The vein is now ready for catheterization (see *Protocol 2*).

Protocol 2. Leukocyte isolation from liver

1. Set up peristaltic pump with polyethylene tubing (0.58 mm i.d., 0.965 mm 0.d.; Intramedic) capable of 2.5 ml/min flow rates.
2. Bevel the free end of the tubing to a 45° angle. Use Hank's buffer containing 10 U/ml heparin as the initial perfusion solution.
3. Start the pump at a 0.2–0.5 ml/min flow rate, and insert the catheter tubing (bevel up) into the hole in the portal vein. Slide it up the vein until it almost reaches the liver; the bevel will now be beyond the ligature.
4. Tighten the ligature to secure the catheter in position. Increase the flow rate to 2.5 ml/min. At this time the abdominal aorta should swell due to increased fluid volume, and it should be severed to allow blood and buffer to escape. The liver should change colour from dark red to tan as the blood is flused from the vasculature.
5. After 5 min, stop the pump and change to the enzymatic digestion buffer. The enzyme digestion buffer is a 20% solution of the enzyme stock in Hank's buffer. The enzyme stock, which should be freshly prepared on the day of use, is 0.25% collagenase, type 1 (Sigma) plus 0.25% dispase, grade II (Boehringer Mannheim, Indianapolis) in Hank's buffer. This should be kept at 4 °C, but it is added to warm (37 °C) Hank's buffer just prior to use. Restart the pump and continue the perfusion for 10 min. When the perfusion is completed, stop the pump and remove the catheter.
6. Carefully remove the liver from the abdominal cavity and separate it from the gall bladder.
7. Transfer the liver into a 60-mm petri dish (on ice) containing Hank's buffer and 100 U/ml DNase type 1 (Sigma).
8. Tease the liver apart with forceps and incubate it for 20 min.
9. Filter the cell suspension through a nylon mesh, and pellet the cells in a centrifuge. This pellet contains both leukocytes and hepatocytes. Separate these from each other by suspending 0.75 ml of the cell pellet in 2.8 ml of 30% Metrizamide (Sigma) in Hank's buffer.
10. Overlay this suspension with 2 ml of PBS and centrifuge at 1500 g for 20–25 min at 4 °C. The leukocytes are at the PBS/metrizamide interface and can be harvested with a Pasteur pipette.

This perfusion technique can also be used to isolate leukocytes from the lungs and other organs.

4.1 Purification of NK cells

A variety of procedures have been published for the purification of NK cells. Mouse NK cells have characteristically been more difficult to purify than human or rat NK cells, and, in general, it is easier to purify NK cells from mice that have first been stimulated with a virus or other IFN inducer, such as polyinosinic:polycytidylic acid (poly I:C). This is because these agents stimulate an increase in NK cell numbers and cell size.

4.2 Purification of endogenous spleen NK cells

Spleen leukocytes are passed through columns of nylon wool (11), which eliminates B cells, granulocytes, and macrophages. Residual macrophages can be removed by plastic adherence, and a variety of monoclonal antibodies and complement (J11d for B cells and granulocytes; lyt 2, L3T4, or CD3 for T cells; B23.1 for macrophages) can be used to further purify the population (see 12). The cells can be sedimented through gradients of percoll and RPMI to collect the band containing low density lymphocytes. We generally use step gradients of 56.6, 52.1, 47.6, 43.1, and 38.6% percoll in complete RPMI medium in a 15-ml conical centrifuge tube. DNase-treated cells are loaded on to the gradient, which is centrifuged at 300 g for 45 min at 20 °C. Endogenous nylon wool-passed NK cells can be collected from the 47.6/52.1% interface (13). Nylon adherent NK cells and activated NK cells band at less dense fractions. If SCID mice, which lack most T and B cells, are available to the investigator, a rapid purification can be achieved by treating the spleen cells simply by antibodies B23.1 and J11d plus complement.

4.3 Purification of activated spleen cells

Stimulation of mice with viruses or other IFN inducers elicits NK cell proliferation as well as accumulation in body organs in which the stimulus is localized. Many techniques can be used to purify stimulated spleen NK cells, and the techniques which we find most useful take advantage of the fact that many of the IFN-stimulated spleen NK cells are large, blast-sized cells. Adult 5–10 week-old mice are injected intraperitoneally with 0.1 mg of poly I:C (Sigma) in 0.1 ml of PBS (a virus or purified IFN can also be used) (14). Three days postinjection the spleens are harvested and processed as described in Section 4. The spleen leukocytes are next treated with DNase for 20 min at room temperature to remove DNA clumps (which also can be removed manually, using a pipette). These cells are next subjected to a procedure which purifies large leukocytes. We use the Beckman centrifugal elutriation system to purify the blast-size cells. Cells are loaded into the system at an r.p.m. setting of 3200 and a medium flow rate less than 15 ml/min. After loading, the flow rate is adjusted to 28 ml/min to elute the small and medium-sized cells. The blast cells are then collected at a flow rate of 46 ml/min. The eluted cells are pelleted and treated with J11d and B23.1 antibodies and complement. These cells in a volume of 2 ml are next sedimented over a discontinuous gradient of 38% (3 ml) and 54% (3 ml) percoll at 2000 r.p.m. for 30 min at 20 °C. The resulting yield is about 85% pure NK cells (about 10^7 cells per 20 mice) (14).

4.4 Purification of activated peritoneal NK cells

The peritoneal cavity normally contains a very low number of NK cells, but their numbers can be greatly enhanced by an intraperitoneal stimulus such as a virus or other IFN inducer. The best reagent in our hands has been mouse hepatitis virus (MHV), strain A59. This is a potent NK cell but poor CTL inducer. Mice are injected i.p. with about 10⁶ plaque forming units of MHV. At 3 days postinfection, the peritoneal cells are harvested as in Section 4, treated to remove erythrocytes, and decanted into a plastic flask or petri dish to allow for macrophage adherence (macrophages represent a major contaminating cell population in peritoneal exudates). After 2 h at 37 °C, the non-adherent cells are removed and treated with J11d plus complement. This will yield a population of NK cells and T cells (15). Addition of antibodies to lyt 2 (CD 8) and L3T4 (CD 4) to the J11d will serve to eliminate T cells, yielding a population highly enriched for NK cells.

4.5 Purification of NK cells by flow cytometry

Thusfar an antibody exquisitely specific for NK cells has not been found in the mouse, but the monoclonal NK 1.1 (16) is the best currently available and has been used by several laboratories to purify and analyse NK cells (12). NK cells from virtually any source can be exposed to this antibody directly conjugated to fluorescein isothicyanate (FITC) or to the unlabelled antibody followed by an FITC anti-mouse IgG isotype-specific reagent. These are examined in a fluorescence-activated cell sorter.

5. Modulation of NK cell activity and number *in vivo*

It is possible selectively to deplete or markedly augment NK cell activity and number *in vivo*, and this is important for examining the role of NK cells during infection. Enhanced NK cell activity and number can be induced by any of a variety of agents, including viruses, bacteria, IFN-β (10⁵ units), poly I:C (0.1 mg), and IL-2 (10⁵–10⁶ units). Intraperitoneal injections tend to produce high levels of NK cells in the peritoneal cavity as well as other organs, whereas intravenous injections result in systemic activation of NK cells in most body parts but not in the peritoneal cavity. NK cell levels usually peak 1–2 days after the peak in IFN induced by the IFN-inducing agents.

5.1 Depletion of NK cell activity *in vivo*

Several reagents are now available for depleting NK cell activity *in vivo*. These include non-specific immunosuppressive drugs such as cyclophosphamide, cyclosporine, and cortisone, but the most selective way to deplete NK cells is with either antiserum to asialo GM1 or with the monoclonal antibody NK 1.1(16).

Both of the reagents can be non-specifically cytotoxic at high concentrations, so they must be titrated downward to the minimum dose required to eliminate the NK cell response. We find that this is usually in the range of 10–40 ul/mouse with the commercially available anti-asialo GM1 reagent (Wako Chemicals, Dallas, TX). The monoclonal antibody to NK 1.1, now available from the American Tissue Culture Collection, is best grown as an ascites. The antibody is precipitated with an equal volume of saturated NH4SO4 and resuspended in PBS at 1 mg/ml. Depending on the preparation, 5–50 μl may be sufficient to deplete NK cell activity.

Cells expressing NK cell antigens increase in number following virus infection *in vivo*, so the dose of antibody required to eliminate NK cells during infection will be higher than that required to eliminate NK cells before infection. Using reagents such as antibody to asialo GM1 at this time is fraught with difficulties in that the high doses of antibody may affect other immune effector mechanisms (17).

6. Demonstrating the anti-viral roles for NK cells *in vivo*

No single method is foolproof to determine the anti-viral roles of NK cells *in vivo*, but a variety of approaches can be used in combination to yield sufficiently interpretable results. One potential problem is that there are probably several mechanisms involved in natural resistance to virus infection and that elimination of one of these mechanisms may not necessarily alter the outcome of infection even though it would if another mechanism were not available. For example, the levels of NK cell activity and IFN usually correlate, but IFN by itself can be a potent mediator of resistance to viral infections. Most of the following experiments can be used to determine whether NK cells are necessary for resistance but cannot be used to show that NK cells never play a role in resistance, because other mechanisms may be compensatory in the absence of NK cells.

6.1 Progression of infection under conditions of NK cell depletion

One of the best ways to quickly assess the importance of NK cells in infection is selectively to deplete NK activity by the anti-NK cell antibody techniques described above in Section 5.1. Mice treated 6 h previously with antibody are then inoculated i.p. or i.v. with virus (this technique has not yet been shown to be very effective by other routes of inoculation). Two to 3 days postinfection, the mice are sacrificed, and their organs are titrated for virus infectivity. Tissue samples can be collected for histological analyses. The advantage of examining these early timepoints after infection is that the specific antibody and T cell immune mechanisms have not yet developed and that one needs to worry less about the effects of NK cell depletion on other immune functions. Titrating the virus is also a much more sensitive way to determine an effect than is a lethal dose assay. We, for example, have found that a given dose of antibody to asialo GM1 which caused a 1000-fold increase in virus titre resulted in only a 4-fold difference in LD50 (18). The dose of virus is, of course, important in these studies, as it is likely that high doses will overcome the resistance provided by NK cells. Care must be taken to examine viral replication at a less than overwhelming dose.

6.2 Progression of virus infection in NK cell-deficient mice

The levels of NK cell activity in mice is under genetic control, and some mouse strains (C3H, CBA) have genetically high NK activity, some (C57BL) moderate activity, and some (BALB, A) low activity. The beige mutant, which is available on several backgrounds, in particular C57BL/6, is a point mutant resulting in a broad-spectrum lysosomal defect which results in low NK cell activity (19). NK-sensitive viruses should grow well in these NK-deficient beige mice.

6.3 Adoptive transfers of NK cells into NK cell-deficient mice

The transfer of 5×10^7 spleen cells from 4–10 week old adult mice into 4–6 day old suckling mice protects the recipients from NK-sensitive viruses (20). The mice should be inoculated with challenge virus within 24 h after the adoptive transfer of leukocytes. Organs from the recipient mice can be titrated 2–3 days later, or else survival curves can be run, in which the frequency and average day of death is determined. To determine whether NK cells are important in the protective effect, two approaches can be made. The first is to use NK cells purified by any of the above listed means, and the second is to deplete NK cell activity from the donor cells either by treating donor mice with antibodies to asialo GM1 or NK 1.1 before harvest of the spleens, or else by treating donor cells with these antibodies plus complement.

7. Generation of lymphokine activated killer cells

Lymphokine activated killer (LAK) cells are culture-generated effector cells capable of mediating NK-like cytotoxicity. A significant subpopulation of LAK cells bears the NK cell phenotype, and these cells can be used in the adoptive transfer systems listed above to provide resistance to certain viral infections. To generate LAK cells, spleen leukocytes are cultured for 5 days at an initial concentration of 3×10^6 cells/ml in RPMI 1640 assay medium containing 5×10^5 M 2-mercaptoethanol, 0.1 μM sodium pyruvate, 0.1 mM non-essential amino acids (M. A. Bioproducts, Walkersville, MD) and 100–1000 units of human recombinant IL-2. A unit from one source of IL-2 may be different from a unit from another source, so it is best to titrate the IL-2 to find an optimum concentration. After 5 days these cells can be tested in cytotoxicity assays or used for adoptive transfer studies.

8. Analyses of steps in the NK cell lytic cycle

Virus infections may enhance or inhibit the susceptibility of cells to lysis by NK cells, and techniques are now available to determine which point in the lytic pathway is affected.

8.1 Binding

For binding studies it is best to use either purified NK cells or else nylon wool-passed spleen cells; otherwise many non-NK cells may bind to targets. Mix purified NK cells with target cells at a ratio of 1:1; mix nylon-passed non-purified NK cells with targets at a ratio of 10:1. These mixtures should be pelleted at low speed and allowed to sit in the pellet for 30 min at room temperature. After gentle resuspension, the cells are examined under a light microscope for effector cell/target cell doublets.

8.2 Calcium flux

The calcium flux, which is an indicator of the transduction of an activation signal across the NK cell membrane, can be measured by two methods. The first method is by the uptake of $^{45}CaCl_2$ into the NK cells (21). Effector cell/target cell doublets in 250 μl are incubated with 5 μCi of isotope at 37 °C for 30 min. The incubation is stopped by the addition of 3 ml of cold medium containing 2 mM lanthanum chloride. The cells are centrifuge-washed in the cold and counted in a scintillation counter for uptake of label. Separate preparations of effector cells and target cells are run as controls.

The second method to study calcium flux is to load effector cells with a fluorescent quinoline Ca^{2+} indicator, quin-2. This binds to free cytoplasmic Ca^{2+} and emits a strong fluorescence signal in the process. The reader is refered to (22) for further details.

8.3 Reorientation of golgi apparatus

After the binding event, NK cells reorient their cytoplasms such that their nuclei become distal to the target cell and their golgi are located between the nuclei and the target. Effector cell/target cell doublets pelleted on to microscope slides are fixed in 4% paraformaldehyde for 8 min and then permeabilized by a 4-min treatment with 0.05% Nonidet P-40 at room temperature. The cells are then stained with fluorescein isothiocyanate-conjugated wheatgerm agglutinin, which will demostrate the polarization of the effector cells (21).

8.4 Secretion

The secretion of granules can best be measured by the detection of a serine esterase localized in the granules. Here, effector and target cells are incubated for 3–4 h in an assay medium containing 1% bovine serum albumin but lacking the phenol red indicator dye and the fetal calf serum. After incubation, the cells are pelleted and 100 μl of the supernatant from this mixture is added to 100 μl of serine esterase buffer. This consists of 0.2 M Tris–HCl, pH 8.1, 0.22 μM Ellmen's reagent (5,5'-dithio-bis-2-nitrobenzoic acid; Sigma), and 0.2 μM BLT (N-benzyloxycarbonyl-L lysine thiobenzyl ester; Calbiochem, La Jolla, CA). The

BLT should be added to the Ellmen's reagent just prior to use. After 30 min incubation at room temperature, the optical density is read at a wavelength of 410.

8.5 Lysis

The lysis of the target cell can be measured by ^{51}Cr release assay or by single cell assay. The single cell assay is particularly useful in these studies, because it will give an indication of what proportion of binding events leads to lysis of the target cell.

9. Measurement of cell migration

NK cells will migrate in response to chemoattractants (15). A very simple method for studying this is to use Boyden chambers, e.g. Nucleopore blind-well chemotactic chambers (Nucleopore, Pleasanton, CA) (15). These chambers contain an upper and a lower portion that can be separated by filters of various pore sizes. A presumptive chemotactic agent in a volume of 0.2 ml is placed in the lower chamber, and a similar volume containing a source of NK cells is placed in the top chamber. Filters are placed between the chambers, which are allowed to incubate at 37 °C for 6–7 h. We have found this time to be optimal, as there is little migration in shorter assays, and longer assays often blur the distinctions between chemotaxis and chemokinesis, which is a non-directed movement of cells exposed to certain activation agents. Two adjacent filters are used in these assays. If the top filter has a 5-μM pore size and the bottom a 0.4 μM pore size, the cells will migrate through the top filter and stick to the impermeable second filter. This second filter can then be stained with Wright's Giemsa, and the cells can be counted. This technique is very easy, but it is sometimes difficult to do differential counts. Cells in chemotaxis assays can be recovered for definitive differential counts and for other analyses if two 5-μM pore-size filters are used. In this case the cells will drop into the bottom chamber, where they can be viably recovered for further analyses. Chemotaxis assays can also be done with radiolabelled cells. The advantage of using the radiolabel is the ease of quantitation of total cell migration; the disadvantage is that, unless the assay is begun with a highly purified cell population, it will not be possible to ascertain which cell types migrated into the bottom chamber.

Acknowledgements

Work in this laboratory is supported by USPHS grants AI17672, CA34461, and AR35506. I thank my former colleagues Drs Christine Biron, Kim McIntyre, and Robert Natuk, for developing some of the discussed techniques.

References

1. Welsh, R. M. (1984). *CRC Crit. Rev. Immunol.* **5**, 55.
2. Welsh, R. M. (1986). *Natural Immun. Cell Growth Regul.* **5**, 169.
3. Hercend, T. and Schmidt, R. E. (1988). *Immunol. Today* **9**, 291.
4. Herberman, R. B., Reynolds, C. W., and Ortaldo, J. R. (1986). *Annu. Rev. Immunol.* **4**, 651.
5. Cerottini, J. C., Engers, H. D., MacDonald, H. R., and Brunner, K. T. (1974). *J. Exp. Med.* **140**, 703.
6. Pross, H. F., Baines, M. G., Rubin, P., Shragge, P., and Patterson, M. S. (1981). *J. Clin. Immunol.* **1**, 151.
7. Hansson, M., Karre, K., Kiessling, R., Roder, J., Andersson, B., and Hayry, P. (1979). *J. Immunology*, **123**, 765.
8. Grimm, E. and Bonavida, B. (1979). *J. Immunology.* **123**, 2861.
9. Biron, C. A. and Welsh, R. M. (1982). *J. Immunology*, **129**, 2788.
10. McIntyre, K. W. and Welsh, R. M. (1986). *J. Exp. Med.*, **164**, 1667.
11. Julius, M. H., Simpson, E., and Herzenberg, L. A. (1973). *Eur. J. Immunol.* **3**, 645.
12. Biron, C. A., Van den Elsen, P., Tutt, M. M., Medveczky, P., Kumar, V., and Terhorst, C. (1987). *J. Immunology*, **139**, 1704.
13. Biron, C. A., Turgiss, L. R., and Welsh, R. M. (1983). *J. Immunology*, **131**, 1539.
14. Biron, C. A., Pedersen, K. F., and Welsh, R. M. (1986). *J. Immunology*, **137**, 463.
15. Natuk, R. J. and Welsh, R. M. (1987). *J. Immunology*, **138**, 877.
16. Seaman, W. E., Sleisenger, M., Eriksson, E., and Koo, G. C. (1987). *J. Immunology*, **138**, 4539.
17. Yang, H., Yogeeswaran, G., Bukowski, J. F., and Welsh, R. M. (1985). *Natural Immun. Cell Growth Regul.* **4**, 21.
18. Bukowski, J. F., Woda, B. A., and Welsh, R. M. (1984). *J. Virology*, **52**, 119.
19. Roder, J. and Duwe, A. (1979). *Nature*, **278**, 451.
20. Bukowski, J. F., Warner, J. R., Dennert, G. and Welsh, R. M. (1985). *J. Exp. Med.*, **161**, 40.
21. Gronberg, A., Ferm, M. T., Ng, J., Reynolds, C. W., and Ortaldo, J. R. (1988). *J. Immunology*, **140**, 4397.
22. Tsien, R. Y., Pozzan, T., and Rink, T. J. (1982). *J. Cell Biol.*, **94**, 325.

6

Viral–receptor binding assays

PETER L. SCHWIMMBECK, MATTHIAS LÖHR, and
ANTOINETTE TISHON

1. Introduction

Viral receptors are the main determinants for viral tropism. Receptors govern pathogenesis by allowing a cell to be susceptible by its attachment to virus. The primary function of receptor proteins on cells is, of course, not the promotion of an infection but a function in the life cycle of the cell. Viruses may utilize several specific host surface proteins like CD4 (human immunodeficiency virus), the Cr2 component of C3 (Epstein Barr Virus), ICAM-1 (rhinovirus), a new member of the immunoglobulin superfamily (poliomyelitisvirus), the adrenergic-like receptor (reovirus) or several proteins of defined molecular weights but not functionally defined (coronaviruses, lymphocytic choriomeningitis virus, etc). Restricted presence of the receptor molecules may account for the tissue specificity along with virus-specific enhancer sequences (1–4). In addition, unrelated viruses may share receptors on a given cell (5).

With the identification of viral receptors they represent a possible target for the modulation or prevention of viral entry into cells. Possible approaches to an intervention are monoclonal antibodies blocking the receptor, synthetic peptides mimicking the viral receptor and thus competing with the host receptor for the attachment of the virus, or drug design of compounds to fit into the receptor groove (9–12).

Two approaches have recently been utilized by our group to detect virus–receptor interactions. One approach is to isolate the membranes containing the viral receptor protein and test it directly for viral binding after immobilization on nitrocellulose membranes. The other approach uses permissive cells and tests the ability of labelled virus to bind to them in solution.

2. Binding to immobilized receptor

2.1. Permissive cells

Select a cell that is permissive for the virus. At a multiplicity of infection of 3, more than 95% of cells should be infectable. This can be detected by immunofluorescence as shown in *Protocol 1*.

Protocol 1. Detection by immunofluorescence

1. Obtain cells either from suspension culture or, if in monolayer, place in suspension by trypsinization using 1% trypsin–EDTA 2–5 min at 37 °C.
2. Wash the cells twice with medium and count them.
3. Resuspend the cells in a small volume (~ 1 ml) and incubate them with the virus at a multiplicity of infection (m.o.i.) of 3.
4. Rotate end-over-end for 1 h at 37 °C.
5. Dilute infected cells to a concentration of 1×10^5 cells/ml. Seed into 24-well tissue culture plates or on coverslips in culture plates. Culture for 24 h.
6. Remove medium and wash three times with PBS.
7. Fix cells with absolute ether/95% ethanol (50/50 v/v) for 10 min, and subsequently with ethanol for 20 min (fixation of cells is done at room temperature).
8. Wash the cells three times with PBS.
9. Incubate the cells with an antibody directed against the virus in a suitable dilution of PBS/10% heat-inactivated FCS for 20 min.
10. Wash the cells three times with PBS.
11. Incubate the cells with FITC-labelled antibody directed against the first antibody in a suitable concentration for 20 min in the dark.
12. Wash the wells and mount coverslips on to glass slides (upside down) with 90% glycerol/10% PBS.
13. Determine the percentage of cells containing viral antigen using a fluorescence microscope.

As controls, uninfected cells should be treated and tested in parallel.

2.2. Preparation of membrane-enriched fractions

The procedure for the preparation of membrane-enriched fractions is as follows in *Protocol 2*.

Protocol 2. Preparation of membrane-enriched fractions (*Figure 1*)

1. Use about 1×10^8 tissue culture cells or 4×10^8 splenocytes for starting material.
2. Grow the tissue culture cells to confluent monolayers, remove the supernatant and wash the flasks $2 \times$ with minimal K Hepes (MKH) buffer, pH 7.4.
    - KCl   130 mM
    - BSA   0.5%
    - Hepes (pH 7)   40 mM

Protocol 2. *Continued*

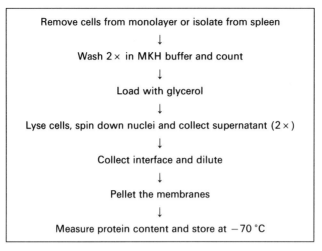

Figure 1. Preparation of plasma membranes from cells

3. Mobilize the cells with a cell scraper and resuspend them in MKH buffer. We avoid using trypsin in order to leave membrane proteins intact.
4. Wash the cells 2 × in MKH buffer and count them. Alternatively, isolate the splenocytes from the spleen by pressing the spleen gently through a steel mesh and subsequently lyse the red blood cells with hypotonic buffer.
5. For the preparation of the membranes, load the cells with glycerol at increasing concentrations of 10%, 20%, and 50% glycerol in MKH buffer (w/v). At the first two steps, incubate the cells for 5 min at 37 °C after addition of glycerol, and after the third, incubate for 10 min. Following this incubation, keep the cells on ice for 5 min and carry out the remaining procedure at 4 °C.
6. Centrifuge at 1200 r.p.m. for 10 min, then lyse the cells by resuspension in disruption buffer for 5 min:
   - Tris–HCl   100.0 mM
   - $MgCl_2$ 1.0 mM
   - $CaCl_2$ 1.0 mM
   - Aprotinin (Calbiochem)   100 U/ml
7. Pellet the remaining intact cells and nuclei by centrifugation at 1500 r.p.m. (about 500 g) for 10 min.
8. Save the supernatant containing the membranes and subject the pellet to a second cycle of cell lysis by incubation in lysis buffer and centrifugation.
9. Pool this supernatant with the previous one and layer on a gradient of 38% sucrose in lysis buffer.
10. Centrifuge at 27 500 r.p.m. for 90 min at 4 °C in a SW41 rotor, then collect

Protocol 2. *Continued*

    the interface, dilute it with lysis buffer and centrifuge it again at 32 500 r.p.m. (about 10 000 g) for 60 min at 4 °C in a SW41 rotor.

11. Collect the pellets by resuspension in 0.01 M Tris–HCl; aliquot and store at −70 °C.
12. Determine the protein concentration of the samples by a standard Lowry assay.

In our experiments, the yield of a preparation was about 1–4 mg of membranes.

In later experiments, to discover whether the putative viral receptor molecule is glycosylated or not and whether unglycosylated protein is still recognized by the virus, cells were grown in the presence of tunicamycin to inhibit N-glycosylation (9, 10) in the biosynthesis of lipid-linked oligosaccharide precursors. Increasing concentrations of tunicamycin up to 2 μg/ml were used and membranes isolated as described above.

Samples of the membrane fractions can be tested *in vitro* with endoglycosidase F (endo-F), that cleaves high mannose structures (9, 10).

2.3 SDS-gel

Separation of the membrane proteins is obtained by SDS-gel electrophoresis (5). The concentration of acrylamide used is dependent on the molecular weight of the membrane proteins studied. The higher the molecular weight of the proteins, the lower the concentration of acrylamide used in the gel. As a rule 10% gels are initially used and then adjusted accordingly.

Protocol 3. SDS-gel electrophoresis

1. Use a vertical electrophoresis system using glass plates in the dimension of 14 × 16 cm. Before use, clear the plates thoroughly with glass cleaner and 95% alcohol to remove all lipids. It is essential to wear gloves during the procedure to avoid contamination of plates.
2. The composition of the stock acrylamide solution (30%) is as follows:
   - acrylamide   30.0 g
   - bisacrylamide   0.8 g
   - H$_2$O   add 100.0 ml
3. The recipes for the separating gel and stacker gel are given below:

	7.5% separating gel	3% stacker gel
30% acrylamide stock	7.5 ml	0.75 ml
1 M Tris, pH 8.8	11.3 ml	–
1 M Tris, ph 6.8	–	1.24 ml
H$_2$O	11.3 ml	7.80 ml
10% SDS	0.3 ml	100 μl

Protocol 3. *Continued*

- 10% ammonium persulphate (fresh)   100 µl   75 µl
- TEMED   15 µl   12 µl
- Total Volume   30.0 ml   10.0 ml

4. Place the plates in the apparatus, then carefully add the separating gel with a 10-ml pipette in order to avoid trapping air bubbles in the gel. Fill the plates to a height of about 4 cm below the top of the glass plates. Allow the acrylamide to solidify (about 30 min).

5. In order to prevent gel drying at the top, overlay the gel with 0.1% SDS in H_2O. After gel solidification, pour off the SDS/water solution, gently pipette the stacker gel on top of the separating gel to avoid bubbles, and put a comb into position.

6. After solidification of the stacker gel (about 20 min), carefully remove the comb.

7. Place the plates with the gel in the electrophoresis chamber and fill both lower and upper reservoirs with running buffer. It is essential to make sure that the seal of the upper buffer reservoir is tight. The composition of running buffer (10 × concentrate) is as follows:
   - Tris base   30 g
   - Glycine   144 g
   - H_2O add to 1 litre

   For use, dilute 1/10 and add 10 ml of 10% SDS per litre of running buffer.

8. Mix aliquots of 100 µg of the membrane preparations with sample buffer to a total volume of about 20 µl, in accordance with the comb size used.
   Sample buffer (3 × concentrate):
   - Tris 1 M, pH 6.8   1 ml
   - SDS 10%   4 ml
   - 2-Mercaptoethanol   1 ml
   - Glycerol   2 ml
   - Bromophenol blue   Add a very small amount in order to obtain a medium blue colour

   For loading the gel, use 2 parts protein solution to 1 part sample buffer.

9. Before loading on to the gel, disassemble the proteins and denature by heating at 65 °C for 10 min or by boiling for 3 min. In some cases, it may be useful to disrupt the membranes by short bursts (2 × 30 sec) of an ultrasonic cell disrupter. Run each gel with a separate lane containing a mixture of marker proteins of different sizes (about 10 µg).

10. After samples are loaded, hook up the gel apparatus to the power supply. (If the gel is run at a high buffer pH, most of the proteins are negatively charged and will migrate toward the anode and hence, the positive pole should be placed at the bottom of the gel). Run gels either at constant voltage (150 V)

Protocol 3. Continued

or at constant power (4 W). The gel should not overheat and melt and should be run until the dye front reaches the bottom of the gel (approximately 4–6 h). To investigate whether there are any disulphide bonds in the protein of interest, use reducing (with BME) and non-reducing gel loading buffers (see Step 8 above).

2.4 Transfer to nitrocellulose

To use the membrane proteins as targets for virus binding, they have to be immobilized by transfer on to nitrocellulose membranes (pore size 0.2 μm). If the putative receptor molecule is a large protein of molecular weight greater than 40 kd, then a larger pore size, such as 0.45 μm, can be used. During the entire transfer procedure it is necessary to wear gloves since small amounts of proteins from the fingers can be transferred and cause artefacts.

Protocol 4. Immobilization of membrane proteins by transfer to nitrocellulose

1. To transfer membrane proteins, remove the gel from the glass plates and place it on filter paper. To avoid air bubbles between the various layers, we recommend that each layer be soaked in transfer buffer [240 ml of running buffer (10 × concentrate), 600 ml of methanol, and 2160 ml of milliQ-H_2O] and kept wet throughout the assembly.
2. Overlay the gel with a sheet of nitrocellulose that has previously been soaked in transfer buffer. Place an additional piece of filter paper on top of the nitrocellulose and the entire sandwich between two sponges and into the transfer apparatus.
3. Fill the apparatus with transfer buffer and turn on the cooling system. The transfer should be done at a temperature not exceeding 20 °C in order to prevent denaturation or partial breakdown of proteins. Most of the proteins are negatively charged due to the high pH of the transfer buffer, therefore, the nitrocellulose membrane must be situated between the gel and the anode to assure proper transfer. The transfer may be done either at 250 mA overnight or at 500 mA for 4–5 h. If transfer is done at 500 mA, the system should be precooled. After transfer, cautiously disassemble the sandwich.
4. For use in the virus binding assay, mark the nitrocellulose membrane and cut it in strips according to the appropriate lanes from the gel.
5. To correlate the virus binding to a defined band and a certain protein size, stain a strip of the nitrocellulose containing the marker proteins with Amido black [45.0% methanol, 2.0% acetic acid (glacial), 0.1% Amido black, and 52.9% H_2O] for about 4 min.

Protocol 4. *Continued*

6. Destain with several changes of destain [10% acetic acid (glacial), 10% methanol, and 80% H_2O], then dry nitrocellulose and keep it for reference. The gel itself can be stained with Coomassie Brilliant Blue [1 g of Coomassie Brilliant Blue, 100 ml of acetic acid (glacial), 500 ml of methanol, and 400 ml of distilled H_2O] for about 45 min.
7. Destain for at least 5 h with several changes of destaining buffer, then either dry the gel on filter paper or take its picture. This allows monitoring of the separation of the proteins, an impression of the purity of the bands and their abundance in comparison to the other membrane proteins.

2.5 Virus binding assay

In principle, this procedure follows the standard Western procedure, and is carried out at room temperature.

Protocol 5. Virus binding assay

1. To block unspecific binding, incubate the nitrocellulose strips in blotting buffer (97.8% PBS, 0.2% Tween 20, and 2.0% non-fat dried milk) for 2 h at 4 °C.
2. Rinse the strips with PBS and incubate for 90 min on a rocker with the virus in a concentration of about 1×10^8 p.f.u./ml in PBS + 3% BSA. In our hands it was not necessary to purify the virus, however, if maximal sensitivity is required, such a step is performed.
3. Wash the nitrocellulose strips with blotting buffer three times each for 10 min, and incubate with an antibody specific for the virus used.
4. Incubate the strips with antibody solution on a rocker for 60 min. Follow with three washes using blotting buffer, each for 10 min.
5. Incubate the strips for 60 min with iodinated [^{125}I]protein A of a high specific activity. For our experiments we iodinated the protein A using the standard procedure with chloramine T. The protein A used had a specific activity of about 5 mCi/mg protein A, and was used in a concentration of about 1×10^6 c.p.m./ml blotting buffer.
6. After inoculation, wash the strips extensively for 4 h using several changes of blotting buffer.
7. Dry the strips, seal them into plastic wrap and expose to film for autoradiography. We used either Kodak X-AR or XR-P film together with intensifier screens. The exposure time used depended on the conditions chosen and ranged from 4 to 96 h. When a mouse or rat monoclonal antibody to the virus is used, a second antibody, i.e. rabbit or sheep anti-mouse or rat Ig is used prior to adding [^{125}I]protein A.

Having identified the band that the virus binds to provides a potential candidate for the putative viral binding site, i.e. receptor. The band can be cut out from the SDS-gel and used for the immunization of animals to raise receptor protein-specific antibodies for additional characterization studies. Obviously, binding dilutions and specific blockings are required.

3. Binding to receptor on cells

The following assay has been adopted from that of Inghirami *et al.* (12). Basically, the procedure makes use of the ability to:

(a) Purify virus.

(b) Label virus with biotin and with about 50% or less loss of infectivity.

(c) Bind biotin-labelled virus to cells in suspension.

(d) Add avidin-flourochrome dye.

(e) Quantitate the number of cells and degree of binding by fluorescent activated cell sorter.

Of interest is the ability to use two different fluorochrome dyes. One for marking virus, the other for the subset of cells involved in the binding. This procedure has been exploited for studying virus binding to unique lymphoid subsets (macrophages, B lymphocytes), B lymphocyte subsets, T lymphocytes, T lymphocyte subsets (T helper, cytotoxic, etc.), natural killer cells.

The procedure has been used successfully for Epstein Barr virus, HTLVI, and LCMV. For completeness we will describe the procedure as we have employed it with LCMV.

3.1 Purification of virus

Protocol 6. Procedure for purification of virus

1. Plate BKH cells at 5×10^6 cells per T175 culture flask containing 50 ml of BKH media

   BKH medium
   - D-MEM   800 ml
   - 1 M Hepes   10 ml
   - 100 × penicillin/streptomycin   10 ml
   - 20% glucose   26 ml
   - 2 × tryptose phosphate broth   50 ml
   - Fetal bovine serum (Heat inactivated) 35 ml
   - Calf serum (heat inactivated)   35 ml

2. The next day aspirate off the culture fluid and add LCMV at an m.o.i. of 1 in 5 ml of media.

3. Put the flasks on a platform rocker at 37 °C for 1 h, then aspirate off the

Protocol 6. *Continued*

   inoculum and add 50 ml of media to the flasks. Put them in a 5% CO_2 incubator at 37 °C for 48 h.

4. Collect culture fluid containing virus and clear it of cell debris by centrifugation at 1200 r.p.m. for 30 min.
5. Add to the culture fluid polyethylene glycol (mol. wt. 7000–9000) 5 g per 100 ml, sodium chloride 2.2 g per 100 ml, and allow it to stir for 3 h or overnight at 4 °C, then spin it at 8000 r.p.m. for 30 min.
6. Dissolve the precipitate in TNE buffer (purification buffer, pH 7.4, 0.1 M NaCl, 0.01 M Tris, and 0.001 M EDTA.)
7. Layer it on a discontinuous renografin gradient consisting of 1 ml 50%, 2 ml 40%, and 3 ml 10%, and spin it in a SW41 rotor at 35 K for 90 min.
8. Collect the banded virus at the interface of 40% and 10% and then resuspend it in TNE buffer.
9. Pellet the virus in a SW41 rotor spun at 35 K for 90 min and dilute in carbonate buffer.
10. Carry out protein determination by Lowry assay.

3.2 Biotinylation of virus

Protocol 7. Procedure for biotinylation of virus

1. Incubate purified virus, 1–2 mg protein/ml in carbonate buffer (0.10 M $NaHCO_3$, and 0.10 M NaCl, pH 8.3–9.0), with *N*-hydroxy-succinimido–biotin (Sigma #H1759) which is diluted in dimethylsulphoxide at 1 mg/ml. Use a ratio of 5 parts of virus to one part of biotin.
2. Prepare the biotin reagent fresh each time. Incubate in glass test tubes at room temperature for 2–3 h or overnight at 4 °C (preferably in the dark). Biotinylation should be done in a NaN_3-free environment.
3. Remove free biotin by dialysis against cold PBS.
4. Dialyse for at least 24 h at 4 °C, changing the dialysate three times a day. Dilute aliquots of biotinylated virus in media containing 7% heat inactivated FCS and store at −70 °C until use.

Following biotinylation, the infectivity of the virus should be determined. If there is more than a 50% loss of titre, the preparation should not be used. A dose–response curve may be generated as follows:

   1. Add increasing amounts (0.5–30 μg) of biotinylated virus to 1×10^6 cells, and incubate them on ice for 45 min.

2. Wash cells twice and add avidin–phycoerythrin (Pe).
   3. After incubating on ice for 30 min, wash cells and analyse on fluorescence activated cell sorter (FACS) to detect the amount of biotinylated virus which gives maximum binding of cells.

To determine the concentration of monoclonal antibody needed, use hybridoma supernatants at neat and then 2-fold dilutions. Follow the above-protocol using FITC antibody against immunoglobulins of the monoclonal antibody.

3.3 Binding of virus
The following protocol is used for binding studies.

Protocol 8. Binding of virus

1. Spin 1×10^6 cells at 1000 r.p.m. for 7 min in 12×75 mm test tubes.
2. Add monoclonal antibody (50 µl) + biotinylated virus (50 µl) simultaneously to cells (concentration of Ab and virus determined by dose–response curve).
3. Incubate on ice for 45 min resuspending cells every 10 min.
4. Add 2 ml of cold media (5% heat inactivated FCS) to cells, spin at 1000 r.p.m. for 7 min.

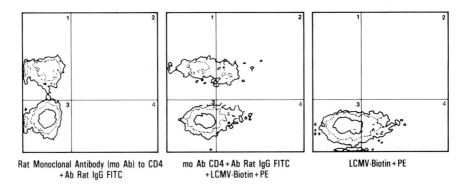

Rat Monoclonal Antibody (mo Ab) to CD4 + Ab Rat IgG FITC   mo Ab CD4 + Ab Rat IgG FITC + LCMV-Biotin + PE   LCMV-Biotin + PE

Figure 2. Demonstration of fluorochrome-labelled virus binding to CD4$^+$ (helper) lymphocytes (Tishon and Oldstone, unpublished observations). Virus (LCMV: lymphocytic choriomeningitis virus) is conjugated to biotin. Ficoll–Hypaque purified lymphocytes are labelled with a monoclonal antibody to CD4 and fluorescein isothiocyanate (FITC). Such lymphocytes are also incubated with LCMV conjugated to biotin and avidin–phycoerythin. When the cells are studied by fluorescence activated cell sorter (FACS) the following results are obtained—**left panel:** CD4$^+$ lymphocytes are moved to frame 1 away from CD4^{nil} lymphocytes in frame 3; **centre panel:** CD4$^+$ lymphocytes (frame 1) that bind LCMV shift to frame 2; **right panel:** total lymphocytes from frame 3 that bind LCMV shift to frame 4. FACS will provide quantitative data as to the numbers or percentage of cells in each frame.

Protocol 8. *Continued*

5. Decant the supernatant and carry out step 4 again.
6. Add FITC-conjugated Ab to immunoglobulins of monoclonal Ab and avidin phycoerythin to pelleted cells.
7. Incubate on ice for 30 min resuspending the cells every 10 min.
8. Repeat steps 4 and 5.
9. For FACS analysis add 1 ml of 0.5–1% formalin in PBS.
10. Run FACS analysis. *Figure 2* demonstrates a successful experiment.

References

1. Mims, C. A. (1986). *J. Immunol.*, **12**, 199.
2. Tignor, G. H., Smith, A. L., and Shope, R. L. (1984). In *Concepts in Viral Pathogenesis*, (ed. A. L. Notkins and M. B. A. Oldstone), p. 109. Springer Verlag, New York.
3. Co, M. S., Fields, B. N., and Greene, M. I. (1986). In *Concepts in Viral Pathogenesis*. (ed. A. L. Notkins and M. B. A. Oldstone), p. 126. Springer Verlag, New York.
4. Boyle, J. F., Weismiller, D. G., and Holmes, K. V. (1987). *J. Virol.*, **61**, 185.
5. Longberg-Holm, K., Crowell, R. L., and Phillipson, L. (1976). *Nature*, **259**, 679.
6. Pert, C. B., Hill, J. M., Ruff, M. R., Berman, R. M. Robey, W. G., Arthur, L. O., Ruscetti, F. W., and Farrar, W. L. (1986). *Proc. Natl. Acad. Sci. USA*, **83**, 9254.
7. Epstein, D. A., Marsh, Y. V., Schreiber, A. B., Newmann, S. R., Todaro, G. J., and Nestor, J. J. (1985). *Nature*, **318**, 663.
8. Rossman, M. G. (1988). *Nature*, **333**, 392.
9. Schwarz, R. T. and Datema, R. (1984). *Methods in Enzymology*, **83**, 432.
10. Elder, J. H. and Alexander, S. (1982). *Proc. Natl. Acad. Sci. USA*, **79**, 4540.
11. Laemmli, U. K. (1970). *Nature*, **227**, 600.
12. Inghirami, G., Nakamura, M., Balow, J. E., Notkins, A. L., and Casali, P. (1988). *J. Virol.*, **62**, 2453.

7

Viral-autoimmune experimental models

ROBERT S. FUJINAMI

1. Introduction

The approach taken for the analysis of virus induced autoimmune disease has been adapted from that for the examination of experimental autoimmune diseases such as allergic encephalomyelitis. The major principle is to use defined viral peptides sharing common amino acids with disease inducing regions from self proteins in order to initiate disease (1). Thus, critical viral determinants which share similar regions with host disease inducing epitopes need to be identified. For our consideration here we will confine the descriptions to the central nervous system (CNS) autoimmune disease and experimental allergic encephalomyelitis (EAE). However, similar approaches can be taken for the study of other organ-specific experimental autoimmune disease such as thyroiditis, diabetes, orchitis and myocarditis.

EAE was one of the first autoimmune diseases described and characterized. It has been a widely used model for the post-infectious encephalopathies and multiple sclerosis. These early studies were initiated to determine the cause of an encephalomyelitis associated with the first rabies vaccines. In testing the vaccines, an autoimmune encephalomyelitis was observed in the 1930s when Rivers and colleagues (2, 3) demonstrated that monkeys repeatedly injected with CNS tissue developed an encephalomyelitis characterized histologically by perivascular infiltrates and demyelination. Clinically, the animals had bouts of incontinence, ataxia and/or paralysis. Animals had immunologically responded to CNS tissue since antibodies to myelin could be detected. However, multiple injections were necessary to produce disease. Finally in 1947, several groups produced an allergic encephalomyelitis in animals using a single injection of CNS material incorporated in Freund's complete adjuvant (4–7). Since then many of the current induction protocols using a variety of CNS antigens have been variations on these pivotal experiments performed in the 1940s (reviewed in 8). One important myelin component, myelin basic protein (MBP) has been identified as an encephalitogenic or EAE inducing protein. This protein has been sequenced and encephalitogenic regions identified for a variety of species (9).

More recently proteolipid protein (PLP), another myelin protein, has been used as an encephalitogen and peptide regions characterized for disease induction in several mouse strains (10). Injection of peptides (encephalitogenic regions) in adjuvant can lead to the development of EAE. Viral peptides having a common sequence with MBP can be substituted for an encephalitogenic site (1). Injection of animals with viral peptides can lead to the development of autoimmunity. This chapter will focus on this aspect.

2. Basic techniques and their utilization

2.1 Identification of common regions

Common determinants between viral and/or various microbial and host cell proteins need to be found and identified. Current technologies in the form of computer software specifically designed for the identification of common nucleic acid and/or protein regions can be used. These programs can be run on commonly available personal computers or with the use of larger mainframe computers. Commercially obtainable communication software is well suited for this purpose and can tie computers into nucleic acid or protein data bases. Alignment and homology searches can be performed to identify sequential determinants. In most instances identical residues searched for in stretches of five amino acids or more an be easily accomplished. Generally, screening in sections of 10 amino acids with overlapping segments of five can be helpful. For most of our studies sequential determinants have been scanned, however, antigenic sites cannot be predicted and common regions may not induce cross-reacting immune responses (11).

2.2 Determination of sites for disease induction

Sites used for the induction of disease have been described for MBP and PLP (12). As yet there is no consistent predictive scheme for selection of disease inducing or encephalitogenic sites. Thus far areas of the molecules have been identified by proteolytic cleavage of the whole molecule and injecting the resulting peptides into the relevant host with adjuvant. The peptide regions can then be further refined through the use of synthetic peptides comprising smaller and smaller sets of amino acids. In this manner the smallest encephalitogenic fragment can be determined.

Thus, once these areas have been identified, they can be used to search against known viral proteins in the commonly used data bases such as Genbank. These regions can then be synthesized and the virally derived peptides used to immunize animals of the appropriate genetic background for the development of disease.

Alternatively, common regions between viruses and host proteins can be first identified by computer analysis and these areas tested for their ability to induce disease directly. The latter is a much less reliable method since disease inducing regions have not been characterized and host disease inducing regions can vary within species as well as from species to species. Care must be taken on interpretation since the production of autoantibodies does not necessarily mean the induction of autoimmunity. Histological and clinical endpoints of disease must also be critically evaluated.

2.3 Peptides

Once the regions have been identified the area in common can be biochemically synthesized. For the most part, the peptides used for study and for the induction of autoimmune disease are not coupled to carriers. Induction of EAE can be achieved with relatively short peptides (reviewed in 9). For example, the induction of EAE in the PLJ mouse requires just an 11 amino acid stretch comprising the amino terminus of the MBP molecule (13).

Peptides can be synthesized using commonly available machines such as that produced by Applied Biosystems (1 and see Chapter 6). The purity of the peptides must be analysed by HPLC to ensure the synthesis was efficient and complete. Very important peptides should be sequenced prior to use.

2.4 Adjuvant

Adjuvants are agents which act non-specifically to enhance immune responsiveness to a specific antigen (14). Animals must be immunized with the peptide in adjuvant. In most instances Freund's complete adjuvant has been used. As mentioned previously this adjuvant has been the one of choice since the advent of experimentally induced autoimmune disease as demonstrated by Jules Freund in the 1940s (4). Freund's complete adjuvant has the ability to enhance both antibody and cellular reactivity (particularly delayed type hypersensitivity responses) to antigens.

The adjuvant performs several tasks which are important for immunogenicity. First, Fruend's complete adjuvant has the ability to induce a local inflammatory response. This allows the accumulation of macrophages and other antigen presenting cells as well as increased expression of class II molecules on these cells. Antigen is then transported to the draining lymph nodes and the immune response is further amplified. Second, mycobacteria contained in the complete adjuvant are a potent macrophage activator. Activated macrophages process and present antigen more efficiently. Third, the adjuvant–antigen emulsion allows for the slow release of antigen. This provides for a more sustained or intense immune response to be generated.

The selected viral peptide in PBS is mixed in the adjuvant containing 1–4 mg/ml of mycobacteria such that an oil in water emulsion is formed. A double-hubbed needle and syringes or a more efficient Sorvall Omnimixer or Tekmar homogenizer can be used to emulsify the antigen viral peptide and adjuvant. The resulting inoculum should be very viscous. For the generation of EAE, rats and guinea pigs can be immunized with less than 5 μg of whole myelin basic protein for the development of disease. In general when using peptide to induce disease larger amounts of peptide antigen are necessary to induce disease (1–10-fold). This may reflect the fact that the whole molecule contains multiple areas which are encephalitogenic while a single peptide contains only one disease inducing site. Similarly, whole CNS myelin induces a better disease than MBP or peptides.

Protocol 1. Sample adjuvant and antigen preparation

Materials
- 1 ml of Freund's Adjuvant incomplete (DIFCO Laboratories, Detroit, MI).
- 4 mg of Mycobacterium tuberculosis H37RA (DIFCO Laboratories, Detroit, MI).
- Antigen (peptide) 5 mg/ml in PBS.

Method
1. Mycobacteria are added to the incomplete adjuvant, and vortexed. This now makes the Freund's adjuvant complete.
2. Add the aqueous phase to the old and homogenize (1 vol to 1 vol) using: Double-hubbed needles and syringes or Sorvall Omnimixer (now Dupont, Claremont, CA) or Tekmar Tissuemizer and microtip (Cincinnati, OH).
3. Final concentration will be 2 mg/ml of mycobacteria, 2.5 mg/ml of antigen. This is the inoculum.
4. Animals are injected with 100 μl of inoculum distributed over several subcutaneous sites.

2.5 Injection scheme

Animals should be injected by the subcutaneous route. Current animal guidelines do not support the use of the footpad route of immunization. For mice, a base of the tail site is recommended. For rabbits, several subcutaneous sites on the flanks allows for the generation of a good immune response. Any additional boosts should be done with Freund's incomplete adjuvant or antigen in alum to prevent necrosis of the original site of injection. Often when using mice, *Bordetella Pertussis* (3×10^9 organisms, Michigan Department of Public Health, Lansing, MI) is given i.v. on 24 and 72 h post immunization as an additional adjuvant boost (15). Animals are followed daily for signs of clinical symptoms.

2.6 Determination of antibody production

At various times after immunization, animals can be bled and sera or plasma tested for antibody production. Several methods have been developed for testing sera or plasma for the detection of autoantibodies.

2.6.1 ELISA

When pure populations of cell-types (i.e. primary astrocytes) can be used, cells are washed twice in PBS. 5×10^4 cells are dispensed into 96-well plates. The plates containing cells are air dried at room temperature. Plates can be stored almost indefinitely (16).

When selective peptide or purified MBP are used as antigen, 100 µl of antigen at 1–10 µg/ml in PBS is added to the wells. The actual concentration should be titrated out. The antigen solution is allowed to adsorb overnight on to the inner surface of the wells at room temperature in humidified chambers. The antigen-coated plates can be stored after the adsorption step by removing the antigen solution and adding 100 µl of PBS containing 0.5 mM thimerosal (see ELISA reagents). Plates can be stored at 4 °C for up to a month when sealed in bags to prevent dehydration.

Next, any non-specifically reactive sites in the wells are blocked with 150 µl of EIA diluent (see ELISA reagents) for 1 h at room temperature. The diluent is then aspirated from the wells and 100 µl of serum dilutions (antibody) added to the wells. The reaction is allowed to proceed for 75–90 min at room temperature in a humidified chamber. The wells are washed three times with ELISA wash (see ELISA reagents) followed by the addition of 100 µl of anti-IgG which is conjugated with horseradish peroxidase. This is followed by another incubation of 75–90 min. The wells are washed three times with ELISA wash and 100 µl of substrate solution is then added. The plates are then incubated in the dark for 30 min. The reaction is stopped by the addition of 100 µl of 1 N HCl. The OD_{492} is read on a Titertek Multiskan photometer.

Protocol 2. ELISA

Sample ELISA

Coating plates

1. Linbro 96-well cluster, (Flow Laboratories, McLean, VA) 100 µl of antigen at 5 µg/ml in PBS.
2. Adsorb overnight at room temperature; keep moist.
3. If need to store, shake off antigen solution, add 100 µl of PBS with 0.5 mM thimersal (Sigma Chemical Co., St. Louis, MO) (a) Refrigerate in a moist chamber or bag.

ELISA

1. Block with 150 µl of diluent, for 1 h, at room temperature.
2. Dilute sera in diluent (see reagents).
3. Incubate for 75–90 min, at room temperature, in a humidified chamber.
4. Wash three times with ELISA wash (gently).
5. Add 100 µl of anti-mouse IgG-HRPO (for the detection of mouse Ig–other second antibodies can be substituted for the detection of other species of Ig) (Tago, Burlingame, CA).
6. Incubate for 75–90 min.
7. Wash and then add 100 µl of substrate solution (see reagents).
8. Incubate for 30 min in dark.
9. Stop the reaction by adding 100 µl of 1 N HCl.

Protocol 2. *Continued*

10. Read OD_{492} on a Titertek multiskan photometer (Flow Laboratories, McLean, VA).

ELISA reagents

Antigens

Usually 5 µg protein/ml is optimum. May need to titrate antigen concentration for optimum results.

EIA diluent
- 90 ml of PBS.
- 10 ml of fetal bovine serum (Flow Laboratories, McLean, VA) or 3% bovine serum albumin (Sigma Chemical Co, St. Louis, MO) final concentration.
- 200 µl of Tween 20 (Sigma Chemical Co, St. Louis MO) final concentration.
- 50 µl of 1 M thimersol (Sigma Chemical Co, St. Louis MO) final concentration.

ELISA wash
- 1000 ml or PBS.
- 2 ml of Tween 20.

Substrate solution
- 20 ml of citrate buffer, pH 5.0
- 8 mg of O phenylene diamine (Sigma Chamical Co, St. Louis, MO).
- 6.6 µl of H_2O_2 (30%).

Citrate buffer
- 5 g of citric acid, anhydrous (Sigma Chemical Co, St. Louis, MO).
- 7 g of Na_2HPO_4, anhydrous (Sigma Chemical Co, St. Louis, MO). Bring up to 500 ml with H_2O and pH 5.0.

1 N HCl
Stock is 12 N.

Thimersol
Final concentration 5 mM.

2.6.2 Immunofluorescence on cells. For the identification of the host or viral components

---

Protocol 3. Immunofluorescence on cells

1. Coverslips (Belco, Vineland, N) in Klean-AR (Mallinckrodt, Paris, KE) for 1 h, then in running H_2O for approximately 1 h.
2. Air dry and place in glass containers (petri dishes), and autoclave.
3. Transport sterile coverslips to 24-well plates (Costar, Cambridge, MA).

Protocol 3. *Continued*

Preparation of cells for placing on to coverslips
1. Trypsinize cells, wash with growth medium and resuspend to a concentration of 1×10^5 cells per ml.
2. Add 1 ml to each plate and place the plates in a 37 °C humidified incubator.
3. Similarly, cells can be grown on commercially prepared multichambered slides (NUNC, Labtek Naperville, IL). Infected cells can also be used.

Fixation of cells
1. After cells have achieved a semi-confluent state, wash the wells twice with warm PBS.
2. Air dry the coverslips and fix in cold acetone (ether–alcohol or other fixatives may be used depending on the stability of the target antigen) for 5 min.
3. Air dry coverslips, after which they can be placed back into dry wells and can be stored frozen up to a month.

Sample assay using mouse serum
1. Thaw and rehydrate coverslips by the addition of 1 ml of PBS to the well.
2. Aspirate the PBS from each well and add 20–40 μl of serum dilution in PBS to the coverslip.
3. Incubate the reaction for 30–60 min at room temperature followed by two washes in PBS.
4. Apply the second antibody (20 μl) anti-mouse Ig fluorescence or rhodamine (Cooper Biomedical Inc., Malvern, PA) labelled to the coverslip. (This must be titrated prior to use.)
5. Allow this binding to proceed for 30–60 min followed by three washes in PBS.
6. Dip the coverslips briefly in distilled H_2O, remove excess H_2O and place cell side down on a glass slide in which a drop of glycerin (Fisher, Fairlawn, NJ)–PBS (9:1) has been applied.
7. Slides view using a UV microscope.

2.6.3 Immunohistochemistry–ABC method staining of paraffin-fixed embedded material

Sections are deparaffinized in Americlear (histology cleaning solvent American Scientific Products, Oakridge, NJ). The sections are then rehydrated in descending concentrations of ethanol (95–30% for 1 min each). Three washes are done with PBS for 5 min each. This is followed by a treatment with proteinase K (ABC reaction) for 15 min at room temperature. To block any non-specific binding the sections are incubated with 5% normal goat serum in PBS for 30 min at room temperature. The first antibody (serum dilution to be tested) diluted in 1% normal goat serum in PBS is added to the sections, and incubated for 2 h at 37 °C in a humidified chamber. After three washes with PBS for 5 min each, the sections are incubated with a 1/50 dilution of biotinylated goat anti-mouse immunoglobulin G (Tago, Burlingame, CA) for 30 min at room temperature, followed by three washes with PBS for 5 min each. The sections are then incubated with an avidin-biotinylated horseradish peroxidase complex as prescribed by the vendor (see ABC reaction). After three additional washes with PBS for 5 min, the sections are treated with diaminobenzidine–hydrogen peroxide solution for 7 min. The reaction is stopped by washing with PBS. After counterstaining with a haematoxylin for 30 sec, the sections are dehydrated in ascending ethanol (30–100% for 1 min each) and coverslipped.

Protocol 4. Sample immunocytochemistry ABC reaction

1. Deparaffinize 3×5 min.
2. Rehydrate in descending ethanol (100–30%) for 1 min each.
3. Wash with PBS 3×5 min.
4. Incubate for 15 min with proteinase K (BRL, Gaithersburg, MD) (1 μg/ml) in 50 mM Tris pH 7.5, 2 mM $CaCl_2$, at room temperature.
5. Block with 5% normal goat serum in PBS for 30 min at room temperature.
6. Dilute the first antibody in 1% normal goat serum in PBS for 2 h at 37 °C
7. Wash with PBS 3×5 min.
8. Incubate with a dilution of 1/50 biotinylated goat anti-mouse, for 30 min at room temperature [during this incubation time prepare A/B complex (Vector Laboratories, Burlingame, CA): 12 μl A + 12 μl B in 1 ml $2 \times$ SSC (17.53 g of NaCl, 8.82 g of sodium citrate, 80 ml of H_2O; adjust pH to 7.0, bring up to 1 litre) (17)].
9. Let the complex form for 30 min at room temperature.
10. Wash with PBS 3×5 min.
11. Treat with diaminobenzidine/H_2O_2 for 7 min at room temperature [dilute 20 mg of diaminobenzidine, Sigma Chemical Co, St Louis, MO, in 100 ml of 0.05 M Tris pH 7.5, filter through Whatman filter paper (Maidstone, England), add H_2O_2 to a final concentration of 0.01%].
12. Stop the reaction with excess PBS.
13. Counterstain with haematoxylin for 30 sec.
14. Dehydrate in ascending ethanol (30–100%) for 1 min each.
15. Coverslip.

2.6.4 Western blotting (immunoblots) for the identification of viral and cellular proteins

Cells and/or tissue are solubilized in SDS sample preparation buffer (see Western reagents). Proteins are separated by SDS gel electrophoresis using standard methods. The proteins are transferred electrophoretically to nitrocellulose paper.

The nitrocellulose is cut into strips and placed in either tubes or petri dishes depending on the size and width of the strips. Enough 3% BSA buffer (see reagents) is added to cover the strips and strips are incubated for 30 min at room temperature. The fluid is then decanted and antibody dilution added. The solution is reacted with the strip for 3 h at 37 °C. The strips are washed three times with 1.5% BSA in PBS. A ^{125}I-labelled second antibody, e.g. goat anti-mouse IG, in 3% BSA in PBS is added ($\sim 1 \times 10^6$ c.p.m.) to the strip. This is incubated for 1 h followed by three washes in 1.5% BSA in PBS. Strips are dried between two paper towels and mounted on cardboard and exposed to X-ray film.

Protocol 5. Sample Western blot assay

After transfer and blocking are complete:

1. Wearing gloves, cut the nitrocellulose paper (Schleicher and Schuell, Keene NH) along the lanes with a scalpel.
2. Depending on the width of the strips, place them in either 5-, 15-ml tubes or petri dishes.
3. Add enough 3% BSA buffer (or 10% instant milk in PBS) to cover the strips, and rotate or shake them.
4. Incubate for 2 h.
5. Add the first antibody (serum dilution)—about 10 ml.
6. Incubate for 3 h at 37 °C.
7. Wash three times (10-min incubation each wash) with 1.5% BSA in PBS.
8. Remove the last wash.
9. Add ^{125}I-labelled second antibody (goat anti-mouse ^{125}IgG) in 3% BSA buffer. 1×10^6 counts/tube or dish is needed (≈ 10 ml).
10. Incubate for 1 h.
11. Wash three times with 1.5% BSA in PBS.
12. Remove the strips using forceps and place them flat on paper towels on an acrylic sheet. Cover the strips with another set of paper towels and acrylic sheet. Change the towels once.
13. Mount the strips on cardboard/stiff paper with tape and expose to X-ray film.

Western blot reagents

3% BSA buffer
- 3% BSA–15 g of bovine serum albumin, Fraction V (Sigma Chemical Co., St. Louis, MO).
- 1% Normal goat serum–5 ml of goat serum (Sigma Chemical Co., St. Louis, MO).
- 0.1% DOC–0.5 g of deoxycholic acid (Calbiochem, La Jolla, CA).

Protocol 5. *Continued*
- 0.5% NP-40–2.5 ml of NP-40 (Sigma Chemical Co., St. Louis, MO).
- 3 ml of 1.0% azide (Sigma Chemical Co., St. Louis, MO).
- Make up to 500 ml with PBS.

$3 \times$ *Sample preparation buffer*
- 30 ml of Solution B
  5.98 g of Trizma base (THAM from Fisher Scientific Co., Fairlawn, NJ).
  0.46 ml of TEMED (Bio-Rad, Richmond, CA).
  Make up to 100 ml with H_2O—use 30 ml.
- 30 ml of 10% SDS (BDH Chemicals Ltd., Poole, UK).
- 30 ml of glycerol (Fisher Scientific Co., Fairlawn, NJ).
- 1.0% 2-Mercaptoethanol (Fisher Scientific Co., Fairlawn, NJ).
- 10 ml of H_2O.

Blocking experiments should be conducted using specific and irrelevant peptides. All the procedures described above can be inihibited using excess antigen (peptide). Several concentrations should be used to achieve a magnitude of antigen excess.

2.6.5 Antibody transfer to determine ability of antibody to transfer disease

Serum from immunized or infected mice is pooled and the Ig concentration determined by ELISA or some other appropriate method. Antibody is then passively transferred into naive recipient animals. Aspects of thryroiditis can be transferred into naive recipients, however, EAE has not convincingly been demonstrated to be transferred by antibody (reviewed in 18). The target organ is then examined histologically for evidence of disease.

2.7 Determination of cellular reactivity

Several methods can be used to determine cellular reactivity. Two methods are described here.

2.7.1 Proliferation assay for mouse spleen cells

This is a conventional technique to determine cellular reactivity to a particular antigen. Two variations can be used. Firstly, animals can be immunized with self antigens and/or peptides and lymphoid cells tested against viral antigens. Secondly, animals can be immunized with viral peptides or infected with virus and responses measured to self determinants.

Spleen from immunized or infected mice are harvested and cleaned of connective tissue. The tissue is gently pushed through a stainless steel mesh and screens washed with medium. The cell suspension is then aspirated up and down several times to break up large clumps and the suspension transferred to a 15-ml conical tube. The larger particles are allowed to settle and the supernatant fluid gently collected. The spleen cell suspension is washed twice and the pellet resuspended in 1 ml of NH_4Cl (0.83%) and incubated at room temperature for 5 min with occasional shaking. Cells are washed twice more in RPMI and resuspended in RPMI containing 5×10^{-5} M mercaptoethanol, vitamins, sodium pyruvate, non-essential amino acids, glutamine and antibiotics with either 10% FCS or 2% normal mouse serum. Cells are dispensed into 96-well plates in 100 μl containing 5×10^5 cells with various concentrations of antigen (peptide or self proteins). Prior to dispensing, cells can be treated with various anti T cells and complement. This can be used to define the T cell subpopulations.

Protocol 6. Sample proliferation assay

1. Push mouse spleen (gently) through a sterile stainless steel screen, using sterile forceps, and scissors.
2. Discard the connective tissues and capsule.
3. Wash the screen with sterile PBS or medium A (see recipe below).
4. Transfer the cells and medium into a 15-ml conical tube and bring up to 10 ml.
5. Allow large particles to settle for 5–10 min.
6. Pipette off the cloudy supernatant containing the cells into a 15-ml conical centrifuge tube.
7. Centrifuge cells at 1000 r.p.m. for 8 min; pour off supernatant fluid.
8. Resuspend in medium A; centrifuge cell suspension again as above.
9. Add NH_4Cl solution (0.83%), approximately 1 ml of NH_4Cl/spleen; resuspend cells.
10. Incubate for 5 min with gentle occasional shaking at room temperature.
11. Add 10 ml of medium A containing 5% FCS; resuspend; centrifuge cells to pellet.
12. Wash the cells once or twice more; resuspend in medium A and count remaining cells.
13. Plate out cells at 4×10^5 cells/100 μl/well in 96-well cluster plates (Costar, Cambridge, MA).
14. Add antigen diluted in medium B (see recipe below), 100 μl/well.
15. At 48 h, pulse with [^{3}H]thymidine, 1 μC/well (Dupont, Boston, MA).
16. At 72 h, harvest with Titertek cell harvester (Flow Laboratories, McLean VA).
17. Count samples.

Protocol 6. Continued

Recipes

Medium A
- 94 ml of 10% FCS–RPMI (with glutamine, antibiotics as usual).
- 2 ml of 5×10^{-3} M solution of mercaptoethanol (Fisher Scientific Co., Fairlawn, NJ)–5×10^{-5} M in final solution.
- 1 ml of 100 × vitamins (Belco Laboratories, Grand Island, NY).
- 1 ml of sodium pyruvate (Irvine Scientific, Santa Ana, CA).
- 1 ml of 100 × non-essential amino acids (Irvine Scientific, Santa Ana, CA).
- 0.5 ml of glutamine (Irvine Scientific, Santa Ana, CA).
- 0.5 ml of antibiotics–Fungi-Bact Solution (Irvine Scientific, Santa Ana, CA).

Medium B
- 92 ml of RPMI [Irvine Scientific, Santa Ana, CA).
- 1 ml of normal mouse serum.
- 0.5 ml L-glutamine (Irvine Scientific, Santa Ana, CA).
- 0.5 ml of 1 M Hepes (Sigma, St. Louis, MO).
- 1.0 ml of Fungi-Bact (Irvine Scientific, Santa Ana, CA).
- 5 ml of 2-Mercaptoethanol diluted 5×10^{-3} M in RPMI (Fisher Scientific Co., Fairlawn, N.J.)–5×10^{-5} M in final solution.

2.7.2 Determination of delayed type hypersensitivity

In EAE, transfer of disease can be mediated by class II restricted cells. Disease induction often correlates with the acquisition of delayed type hypersensitivity to encephalitogenic proteins. Thus, this is a useful parameter to measure.

Generally, on day 21 post inoculation, mice are challenged with the appropriate antigen (peptide) in the left footpad and PBS in the right footpad. Infected mice are injected with 10 µg of inactivated virus in 30 µl of PBS. Control mice immunized with ovalbumin are injected with 30 µl of 5% heat-aggregated ovalbumin. Footpads are measured 4, 24, and 48 h after challenge with a Mitutoyo Model 7309 micrometer caliper. The increase is swelling of the antigen-challenged footpad versus the PBS control footpad is determined, and the results expressed in units of millimeters [19].

2.7.3 Cell transfer to determine cellular aspects of disease induction

To definitively ascertain the immunopathological nature of the disease, cell transfer experiments can be conducted. The first transfer of EAE was performed in 1960 by Paterson (20).

The draining lymph node and spleen from sensitized or virus infected mice are removed and teased in medium. The cell suspension is adjusted to 4×10^6 cells/ml and cultured with antigen at 100 µg/ml in RPMI supplemented with 5×10^{-5} M mercaptoethanol, sodium pyruvate, non-essential amino acids, glutamine and antibiotics. Cells are cultured for 4 days in Linbro 24-well cluster plates at 37 °C.

Afterwards, cells are pooled and washed three times and adjusted to the appropriate concentrations in 100–200 µl for i.v. injection into recipient mice. Mice are checked daily for clinical signs (21). When cells are obtained from infected animals, cautious and critical evaluation must be performed to rule out transfer of virus.

3. Determination of injury or pathology

Examination of the target tissue i.e. CNS for production of injury resulting from immunopathological damage should be conducted. This will determine whether the cross-reacting immune response actually progresses to disease. In many instances auto-antibody can be detected, but no disease results (reviewed in 22).

At various times post transfer of immune reagents (cells and/or antibodies) or immunization with viral peptide, animals should be euthanized, serum collected to detect antibody and lymphoid cells tested for their ability to respond to test antigen. Tissues can be fixed with 4% paraformaldehyde and processed for histological examination. CNS can also be processed for plastic embedding and the extent of demyelination ascertained.

Acknowledgements

This work was supported by grants from the National Institutes of Health and the National Multiple Sclerosis Society. Special thanks to Jan Richards for excellent manuscript preparation.

References

1. Fujinami, R. S. and Oldstone, M. B. A. (1985). *Science*, **230**, 1043.
2. Rivers, R. M. and Schwentker, F. F. (1935). *J. Exp. Med.*, **61**, 689.
3. Rivers, T. M., Sprunt, D. H., and Berry, G. P. (1933). *J. Exp. Med.*, **58**, 39.
4. Freund, J., Stern, E. R., and Pisani, T. M. (1947). *J. Immunol.*, **57**, 179.
5. Morgan, I. (1947). *J. Exp. Med.*, **85**, 131.
6. Kabat, E. A., Wolf, A., and Bezer, A. E. (1947). *J. Exp. Med.*, **85**, 117.
7. Morrison, L. R. (1947). *Arch. Neurol. Psychiatry*, **58**, 117.
8. Paterson, P. Y. ( ). In *Autoimmunity*, (ed. N. Talal), p. 643. Academic Press Inc., New York.
9. Hashim, G. A. (1978). *Immunol. Rev.*, **39**, 60.
10. Tuohy, V. K., Lu, Z., Sobel, R. A., Laursen, R. A., and Lees, M. B. (1989). *J. Immunol.*, **142**, 1523.
11. Dyrberg, T. and Oldstone, M. B. A. (1986). *Curr. Topics Microbiol. Immunol.*, **130**, 25.
12. Lees, M. B. and Macklin, W. B. (1988). In *Neuronal and glial proteins*, (ed. P. J. Marangos, I. C. Campbell, and M. R. Cohen), p. 267. Academic Press Inc., San Diego, Ca.
13. Zamvil, S. S., Mitchell, D. J., Moore, A. C., Kitamura, K., Steinman, L., and Rothbard, J. B. (1986). *Nature*, **324**, 258.

14. Maillard, J. and Bloom, B. R. (1972), *J. Exp. Med.*, **136**, 185.
15. Fritz, R. B., Jen Chou, C.-H., and McFarlin, D. E. (1983). *J. Immunol.*, **130**, 191.
16. Rice, G. P. A. and Fujinami, R. S. (1986). In *Methods of Enzymatic Analysis*. (ed. H. U. Bergmeyer), p. 370. Verlag Chemie.
17. Maniatis, T., Fritsch, E. F., and Sambrook, J. (1982). *Molecular cloning, A laboratory manual.* Cold Spring Harbor Laboratory Press, NY.
18. Weigle, W. O. (1980). *Adv. Immunol.*, **30**, 159.
19. Titus, R. G. and Chiller, J. M. (1981). *J. Immunol. Methods*, **45**, 65.
20. Paterson, P. Y. (1960). *J. Exp. Med.*, **111**, 119.
21. Pettinelli, C. B. and McFarlin, D. E. (1981). *J. Immunol.*, **127**, 142.
22. Fujinami, R. S. (1988). *Ann. NY Acad. Sci., Adv. Neuroimmunol.*, **540**, 210.

8

Synthetic peptides as antigens

THOMAS DYRBERG and HANS KOFOD

1. Introduction

Antibodies have always had a prominent place in the armamentarium available to research. In general, antibodies to complex antigens obtained from immunized animals or through hybridoma technology are characterized by a high degree of specificity; however, the exact position of the binding site on the antigen is only rarely known. Further, the production of such antibodies requires that the antigen is available, either in a purified state or as part of a heterogeneous mixture. One approach to circumvent some of these problems has been to generate antibodies to defined epitopes of proteins, using synthetic peptides, representing sequences of the proteins in question, as antigens (1–3). As the methods to synthesize peptides have been made more accessible, this approach has been widely used to produce peptide antibodies of predetermined specificity which has proved to be helpful in several respects:

(a) In identifying gene products deduced from open reading frames of nucleotide sequences (4).

(b) As probes to differentiate between homologous proteins. Antibodies raised to synthetic peptides differing by only one amino acid can distinguish between closely related proteins (5). This type of antibodies may be of considerable use in analysis of polymorphic proteins, e.g. of MHC molecules characterized by a high degree of polymorphism which depends on only few amino acid substitutions between alleles.

(c) To establish the orientation of proteins in membranes (6) or sites of interaction with other components (7).

(d) To investigate if molecular mimicry, i.e. the presence of shared amino acid sequences between virus and host cell proteins could contribute to the development of autoimmunity (8).

(e) In the development of synthetic peptide vaccines (9) and in particular the capacity of defined protein sequences from pathogenic microorganisms to induce neutralizing antibodies.

Most B-cell epitopes in a protein are conformational, but there are nevertheless ample examples of antibodies raised to short, synthetic peptides that bind to intact or native proteins. The ease with which protein-reactive peptide antibodies are produced has, however, been a matter of debate (2, 10). Immunization of an animal with a peptide, conjugated to a carrier-protein, will likely result in peptide-binding antibodies, often of very high titers. However, the success rate in producing peptide antibodies that react with the parent protein from which the sequence was obtained may vary depending, e.g. on the length of the immunizing peptide, on the choice of carrier-protein and method of conjugating the peptide to the carrier-molecule, on the species of the animal used for immunization and on the method of analysis of the antibodies (theoretically, peptide antibodies should bind more easily to denatured proteins where the epitope will be exposed than to native proteins in their three-dimensional configuration, where the epitope may be less accessible).

During recent years, there have been several extensive reviews on peptide antibodies (1–3) and the practical aspects concerned with this technique (11). The intention with this chapter is not to provide an exhaustive review of the various steps involved in producing peptide antibodies, but rather to describe the methods of peptide synthesis and purification, conjugation to carrier-protein and immunization that we have used in producing antibodies to more than 50 different peptides, representing various cellular proteins.

2. Selection of sequences

Selection of the amino acid sequence for a synthetic peptide antigen may be based on several criteria depending on the use of the antibodies. Principally, the sequence may represent either a specific site of interest in a given protein, e.g. a suspected clevage site in a precursor molecule, a site defining a particular allelic molecule (5), a pathogenic epitope on an autoantigen (12), or the sequence may be selected with the primary goal of inducing antibodies which crossreact with the intact protein, e.g. identification of an unknown gene product of an open reading frame. In the former case, the exact site of interest needs to be known in advance; however, information obtained from using peptides and peptide antibodies can in turn help in identifying such sites (13). The situation is more difficult in the latter case. From analysis of model proteins where the three-dimensional structure is known, the location of antigenic epitopes has been correlated with factors such as hydrophilicity, mobility, and accessibility. By taking these findings into account various algorithms have been claimed to be able to predict the localization of an antigenic epitope based on the primary structure of a protein (2, 14,15). However, the generalization from the structure/antigenicity relationship of model proteins to amino acid sequences of unrelated proteins has in general not been very helpful in selecting antigenic sequences that induce protein crossreactive peptide antibodies. The crossreactivity of a peptide antibody with the intact protein from which the sequence was derived or even with a homologous peptide, has to be established empirically and crossreactivity may depend on factors such as single amino acid substitutions (16) or the orientation of the peptide to the carrier protein (17).

3. Peptide synthesis

Merrifield introduced the concept of solid-phase peptide synthesis in 1963 (review in 18). This method has since been effectively automated and is now widely available. The principal steps involved in the solid-phase peptide synthesis are:

(a) The protected, carboxyl-terminal amino acid is attached to a solid support by a covalent bond (which has to be stable during the synthesis).

(b) The free α-amino group is regenerated in such a way that the side chain protection groups remain unaltered.

(c) The next protected amino acid is added to the first by the formation of an amide band. Steps (b) and (c) are repeated until a peptide of the desired sequence has been synthesized.

(d) At the end of the synthesis, the protected peptide is cleaved from the solid support and the side chain protection groups removed.

4. Purification of peptides

With the available automated peptide synthesizers, peptides are efficiently produced; however, they often have to be purified before use. The purity of the final product may depend on the use of the peptide, e.g. peptides used for functional studies often need to be more purified than peptides used for immunization. In the following we describe a simple method to isolate purified peptide using step gradient elution of the sample from a reverse-phase column. This method has proven to be applicable to a wide range of peptides.

Protocol 1. Peptide purification

Materials
- C18 reverse-phase mini-column (e.g., Sep-Pak, Waters Associates)
- 2% acetic acid (v/v) in water
- acetonitrile

1. Dissolve 30 mg of the impure peptide in 10 ml 2% acetic acid and inject the solution with a syringe through a C18 cartridge.
2. Wash the column with 10 ml of 2% acetic acid.

Protocol 1. Continued

3. Elute the peptide from the cartridge with increasing concentrations of acetonitrile in 2% acetic acid: 10 ml 2.5% acetonitrile, 10 ml 5% acetonitrile, 10 ml 7.5% acetonitrile, etc. until the peptide has been eluted.
4. Examine all the fractions by analytical reverse-phase HPLC (a representative example is shown in *Figure 1*). The peptide is isolated by lyophilization of the fraction(s) containing the purified peptide.

If the purity of the peptide is not satisfactory after these procedures, the fractions containing the peptide can be diluted with 10 ml 2% acetic acid and reapplied to a new C18 cartridge. Elute the peptide in 10 ml volumes of 2% acetic acid with acetonitrile, by using smaller increments of the acetonitrile concentration in the range where the peptide was eluted from the first column. These steps usually result in products with a purity higher than 98% (*Figures 1* and *2*). The final peptide product should be adequately identified before use, preferably by more than one method, e.g. HPLC (*Figure 1*), amino acid analysis, amino acid sequence analysis, and mass spectrometry (*Figure 2*).

5. Peptide conjugation to carrier protein

In order to produce high-titered antibodies to small peptides, these compounds need to be coupled to a larger carrier-molecule. Peptides may be coupled to carrier-proteins at various points of attachment, e.g. through the terminal amino acid residue via either α-amino- or α-carboxyl-groups or through reactive sites in cysteine, tyrosine, histidine, or lysine residues (11). The most commonly used carrier proteins include keyhole limpet hemocyanin, bovine serum albumin, thyroglobulin and tetanus toxoid. The peptide can be coupled to the carrier-protein by homo- or bifunctional reagents which either become incorporated into the final conjugate or activate certain reactive sites of the carier-protein for subsequent linkage with the peptide (11). Below, we describe the method we have used for conjugation of peptides to keyhole limpet hemocyanin, essentially according to the method described by Liu *et al.* (19). The carrier-protein is activated by reaction of the heterofunctional reagent *m*-maleimidobenzoyl-*N*-hydroxysuccinimide ester with the ε-amino group in lysine residues and the α-amino of the amino terminal residue (*Figure 3*). The peptide is subsequently bound to the activated carrier-molecule through the thiol group in a cysteine residue (*Figure 3*). The orientation of the peptide to the carrier is therefore determined by the position of the cysteine residue in the peptide (17). Since the coupling of peptide to the activated carrier-molecule depends on the presence of a reduced thiol group in the cysteine residue, peptides should be assayed for cysteine activity before conjugation (20) (*Protocol 4*).

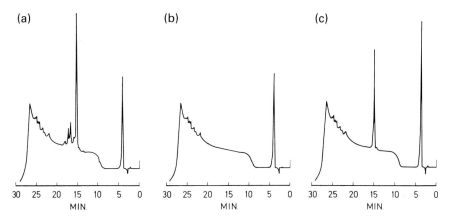

Figure 1. HPLC analysis of the purity of a peptide (PQVPQLELGGGPEAGDLC) synthesized by solid-phase methodology on an Applied Biosystems 430A Peptide Synthesizer and purified by stepwise gradient elution from a C18 mini-column. HPLC-conditions: Solvent speed: 1 ml/min, Column: 5 μm C18 Licrosorp, dimensions: 4 × 250 mm. Solvent A: 0.1% trifluoroacetic acid in water, solvent B: 0.1% trifluoroacetic acid in acetonitrile. Solvent profile: 100% A for 10 min, 100% A to 100% B over 20 min. (a) HPLC profile of the crude peptide after HF-cleavage. The material eluting at 4 min is the solvent peak. (b) The material eluted from the mini-column after washing with 2% acetic acid-water. (c) The pure peptide eluted in 20% acetonitrile in 2% acetic acid-water. The peaks after 20 min represent contaminants in the water. Peptide contaminants (eluting from 16 to 18 min in panel A) are eluted from the mini-column with acetonitrile concentrations above 20% (not shown).

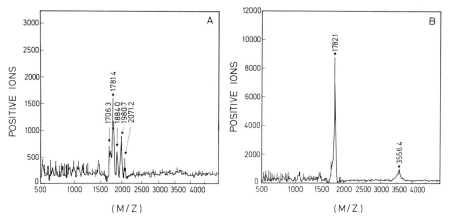

Figure 2. Mass spectrometry of the peptide shown in Figure 1 (recorded on a BIOION-20 Time of Flight instrument). (A) Crude HF-cleaved product. (B) The purified peptide.

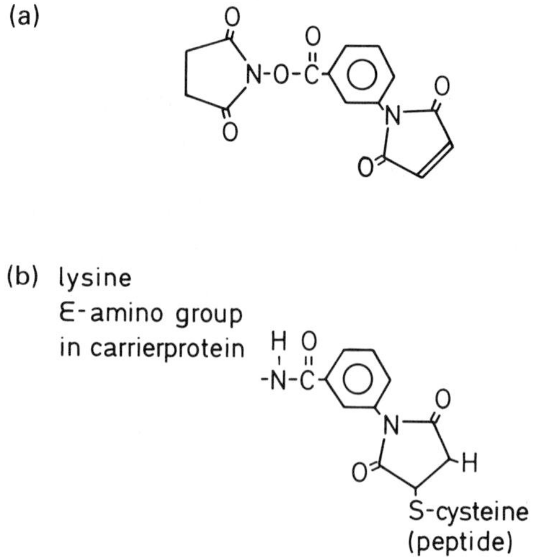

Synthetic peptides as antigens

Figure 3. Activation of carrier-protein with MBS: (a) m-maleimidobenzoic acid is coupled to free $-NH_2$ groups in the carrier protein through the N-hydroxysuccinimide activated carboxylic acid. (b) The peptide antigen is coupled to the activated carrier protein by addition of the free -SH group to the carbon-carbon double bond in the maleic acid moety (lower panel).

Protocol 2. Coupling protocol

Materials
- KLH (keyhole limpet hemocyanine)
- MBS (m-maleimidobenzoyl-*N*-hydroxysuccinimide ester)
- DMF (dimethylformamide)
- 10 mM phosphate buffer, pH 7.2
- 50 mM phosphate buffer, pH 6.0
- G-25 Sephadex

Method
1. Dissolve KLH in 10 mM phosphate buffer, pH 7.2 to a concentration of 15.6 mg/ml and dialyse for 24 h against the same buffer. 5 mg KLH is coupled to 5 mg peptide, this will provide for the immunization of 2–3 rabbits (the ratio of peptide to KLH may be varied; molar ratios between 5 and 50 have been reported to be suitable for immunization).
2. Dissolve MBS in DMF to a concentration of 12 mg/ml immediately before use.
3. To activate KLH, slowly add 55 µl MBS (12 mg/ml) to each volume of 320 µl KLH (5 mg) and incubate for 30 min at room temperature. (The final concentration of DMF in the reaction volume should be below 30%, since DMF causes KLH to come out of solution).

Protocol 2. *Continued*

4. Equilibrate a column of 20–25 ml Sephadex G-25 with 50 mM phosphate buffer, pH 6.0. Apply the activated KLH to the column and collect the eluate in 1 ml fractions. Measure the absorbance at 280 nm–the first peak contains activated KLH, the second MBS. Pool the fractions containing the activated KLH and calculate the concentration (assume 100% recovery).

5. Immediately before use, dissolve 5 mg of peptide in 1 ml of degassed, redistilled water. Add 5 mg of activated KLH to the peptide sample under constant agitation. Adjust pH to 7.0–7.5 with NaOH or HCl and incubate at room temperature under constant agitation for 3 h. The conjugate can be stored at $-20\,°C$.

Protocol 3. Cysteine assay

Standard curve

Sample	Glutathione[a]	PE[b]	DTNB[c]
1	0	1000 µl	100 µl
2	0	1000 µl	100 µl
3	2.5 µl	1000 µl	100 µl
4	5 µl	995 µl	100 µl
5	10 µl	990 µl	100 µl
6	15 µl	985 µl	100 µl
7	20 µl	980 µl	100 µl

[a] *Glutathione*: 1.44 mg/ml in PE.
[b] *PE*: 0.1 M phosphate buffer, 5 mM EDTA, pH 7.2.
[c] *DTNB*: 1 mM dithionitrobenzoic acid in methanol.

The glutathione standard and DTNB may be stored at $-20\,°C$ until use. Dissolve the peptide in degassed, redistilled water immediately before use.

Peptide samples
25 µl sample peptide, 5 mg/ml
975 µl PE
100 µl DTNB

Method

1. Incubate the samples for 30 min at room temperature.
2. Measure the absorbance at 412 nm.
3. Calculate cysteine concentration according to the standard curve.

---

6. Immunization

High affinity peptide antibodies are easily produced by immunizing rabbits or mice with conjugated peptides in adjuvants (*Protocol 5*). The problem is, however, to generate antibodies which crossreact with the intact protein and the choice of carrier-protein, conjugation method, adjuvant and immunization scheme may influence the result (11). Although a peptide antiserum theoretically displays a limited epitopical diversity, the procedure often results in non-specific reactions, e.g. to the carrier-protein which may increase non-specific binding in radioimmunoassays, Western blot analysis or immunocytochemistry. One approach to reduce non-specific binding of a polyclonal, site specific antiserum, is to isolate peptide specific antibodies by affinity chromatography (21). Alternatively, monoclonal peptide antibodies may be produced by hybridoma technology. This will ensure an unlimited supply of antibodies, often with low non-specific binding, however, the binding affinity of a hyper immune serum is seldom obtainable with a monoclonal antibody. Recently, it has been suggested that high-affinity monoclonal antibodies of the IgG isotype are more frequently isolated following a combination of *in vivo* and *in vitro* immunizations (Borrebaeck, personal communication).

Protocol 4. Immunization protocol

Animals
Female rabbits, approx. 2 kg, 2–4 rabbits per peptide.
Antigens
Peptides of 8–20 amino acids, coupled to keyhole limpet hemocyanine (KLH).

Day	Treatment (per rabbit)
–	Bleed for preimmune serum.
0	Emulsify 250 µg peptide, coupled to KLH in an equal volume of complete Freunds adjuvant, total volume 1 ml. Shave the back of the rabbit and inject the peptide emulsion intradermally at 30–50 sites.
14	Emulsify 250 µg peptide, coupled to KLH, in an equal volume of incomplete Freunds adjuvant in a total volume of 0.5 ml. Inject subcutaneously in the back.
21	Immunize as on day 14.
28	Bleed for the 1st postimmune serum.
35	Immunize as on day 14, but use only 150 µg peptide.
42	Bleed for the 2nd postimmune serum.

The rabbits can be immunized and bled for several months according to this scheme.

Acknowledgements

Thomas Dyrberg was supported by a Career Development Award from the Juvenile Diabetes Foundation.

References

1. Lerner, R. A. (1984). *Adv. Immunol.*, **36**, 1.
2. Van Regenmortel, M. H. V. (1989). *Immunol. Today*, **10**, 266.
3. Walter, G. (1986). *J. Immunol. Methods*, **88**, 149.
4. Lerner, R. A., Sutcliffe, J. G., and Shinnick, T. M. (1981). *Cell*, **23**, 309.

5. Atar, D., Dyrberg, T., Michelsen, B., Karlsen, A., Kofod, H., Molvig, J., and Lernmark, Å. (1989). *J. Immunol.*, **143**, 533.
6. Schneider, W. J., Slaughter, C. J., Goldstein, J. L., Anderson, R. G. W., Capra, J. D., and Brown, M. S. (1983). *J. Cell Biol.*, **97**, 1635.
7. Sytkowski, A. J. and Donahue, K. A. (1987). *J. Biol. Chem.*, **262**, 1161.
8. Dyrberg, T. and Oldstone, M. B. A. (1986). *Curr. Top. Microbiol. Immunol.*, **130**, 25.
9. Steward, M. W. and Howard, C. R. (1987). *Immunol. Today*, **8**, 51.
10. Jemmerson, R. and Blankenfeld, R. (1989). *Mol. Immunol.*, **26**, 301.
11. Van Regenmortel, M. H. V., Briand, J. P., Mullers, S., and Plaue, S. (1988). In *Laboratory Techniques in Biochemistry and Molecular Biology*, (ed. R. H. Burdon and P. H. van Knippenberg). Elsevier, Amsterdam.
12. Fujinami, R. S., Oldstone, M. B. A. (1985). *Science*, **230**, 1043.
13. Schwimmbeck, P. L., Dyrberg, T. Drachman, D. B., and Oldstone, M. B. A. (1989). *J. Clin. Invest.*, **84**, 1174.
14. Getzoff, E. D., Tainer, J. A., Lerner, R. A., and Geysen, H. M. (1988). *Adv. Immunol.*, **43**, 1.
15. Van Regenmortel, M. H. V. and Daney de Marcillac, G. (1988). *Immunol. Lett.*, **17**, 95.
16. Alexander, H., Johnson, D. A., Rosen, J., Jarabek, L., Green, N., Weissman, I. L., Lerner, R. A. (1983). *Nature*, **306**, 697.
17. Dyrberg, T. and Oldstone, M. B. A. (1986). *J. Exp. Med.*, **164**, 1344.
18. Barany, G. and Merrifield, R. B. (1979). In *The Peptides*, (ed. E. Gross and J. Meienhofer), Vol. 2, p. 3. Academic Press, New York.
19. Liu, F. T., Zinnecker, M., Hamaoka, T., and Katz, D. H. (1979). *Biochemistry*, **18**, 690.
20. Means, G. E. and Feeney, R. E. (1971). *Chemical modification of proteins*, pp. 155–7. Holden-Day, Inc., San Francisco.
21. Vaughan, J. M., Rivier, J., Corrigan, A. Z., McClintock, R., Campen, C. A., Jolley, D., Vogelmayr, J. K., Bardin, C. W., Rivier, C., and Vale, W. (1989). *Methods Enzymol.*, **168**, 588.

Index

Agarose minigels 44
Antibodies 19, 36
  monoclonal 21–2
  polyclonal 22
Antibody
  nucleic acid probe double-labelling 84–5
  production 152
Antigen–antibody interactions 16
Antigens 163
Autoimmune disease 149

β-galactosidase
  recombinant marker protein 106–7
Biotinylation of virus 145–6

Clq
  radiolabelling 89
  standard curve 90–1
Cell lysis 17–18
Central nervous system (CNS) 149
^{51}Chromium
  labelling of cells 97
Complement
  detection 92
Cryomicrotomy 8–9
CTL clones
  carrying 103
  expanding 103
  freezing and storing 103–4
  generation 99–100
  immunosuppression of recipient host 118
  inoculation 118
  migration studies 118–19
  removal of dead cells 116–17
  use in vivo 104, 115
Cytotoxicity 122–3
Cytotoxic T cells 104, 121
Cytotoxic T lymphocyte (CTL) assay
  evaluation 98–9
  calculation 98–9
Cytotoxic T lymphocytes (CTLs) 104
  ^{51}Chromium assay 95
  major histocompatibility antigen complex 95
  peripheral blood 96–7
  precursor frequency 98
  target cell genetics 95
  target cell selection 95–6
  virus specificity 95

Differentiated cell 1
Direct injury 1
DNA
  amplification 63–4
  extraction 57–60

Electrophoretic transfer 27, 31–3
ELISA 152
Epitope
  CTL 113
  mapping 113
Experimental allergic encephalomyelitis (EAE) 149
Expression of foreign genes in vaccinia virus 108–11

Fluorescence activated cell sorter (FACS) 146
Fluorescence microscopy 92

Gel electrophoresis 30–1
Generation of CTL
  spleen 99
  T cell growth factor 102
  limiting dilution 102–3
Glyoxal 46
Guanidinium thiocyanate (GTC) 40–3

Histology 5
Hybridization probes
  non-isotopically labelled 52–3
  radiolabelled 47

Immune complexes
  body fluids 87
  Clq binding test 87–8
  Raji test 87–8
  serum 87
  tissue deposits 87, 92
Immunochemical probing 34–5
Immunocytochemistry 67, 71
  double-labelling 77
  single-labelling 75–7
Immunoglobulin
  detection 92
Immunologic injury
  antibody-dependent cell mediated 93

Index

Immunologic injury (*cont.*)
   antibody mediated 92
   complement mediated 92
Infectious virus-antibody immune complexes
   detection 91
   presence 91
In situ hybridization 5, 9–10, 67, 77–84
Interferon (IFN) 121

Keyhole limpet hemocyanine (KLH) 168–9

Large granular lymphocyte (LGL) 121
Lymphokine activated killer (LAK) cell 132

Membrane preparation 138–40
Molecular mimicry 149, 163
Monoclonal antibodies 22–3
Myelin basic protein (MBP) 149

Natural killer (NK) cells 121
Non-isotopic hybridization probes 52–3
Nucleic acids
   analysis 60
   isolation 57

Oligo (dT) cellulose 45

Paraffin-embedded tissue 55–6, 155
Pathogenesis 5
Peptide
   conjugation 166
   purification 165–6
   synthetic 114
Peritoneal macrophages
   harvesting 100–2
   use to maintain CTL clones 100–4
Polyclonal antibodies 22
Polymerase chain reaction 60, 64–6
   preparation of primers 62
   selection 62
   sequencing 65

Proliferation assay
   mouse spleen cells 158–60
Protein A 23, 36
Protein blotting 14
pSC_{11}
   cloning into 106–7
Purification of virus 144–5

Radiolabelled proteins 25
Reverse transcription 65
RNA
   extraction 57–60
   transfer 40

SDS-PAGE 140–2
   electrophoresis 28–30
Serine esterase 121
Sequencing 65
Slot blot 60
Synthetic peptides 163

T cell epitopes 104
T cell growth factor
   preparation 102
   use 102
Tissue fixation 68–9

Vaccinia virus recombinants 106–7
Vacuum blotting 52
Viral antigen
   detection 92
Viral receptors 137
Virus binding assay 143–4
Viruses
   alter cell function 1
   cell-cell fusion 1
   indirect injury 2
   perturb membranes 1
   produce toxins 1
   shut down protein synthesis 1

Western blot assay 14–15, 156–8
Whole animal sections 5–14